ECONOMIC-MATHEMATICAL METHODS AND MODELS UNDER UNCERTAINTY

ECONOMIC-MATHEMATICAL METHODS AND MODELS UNDER UNCERTAINTY

A. G. Aliyev, PhD

Apple Academic Press

TORONTO NEW JERSEY

Apple Academic Press Inc. 3333 Mistwell Crescent Oakville, ON L6L 0A2 Canada	Apple Academic Press Inc. 9 Spinnaker Way Waretown, NJ 08758 USA

Exclusive worldwide distribution by CRC Press, a member of Taylor & Francis Group

Printed in the United States of America on acid-free paper

International Standard Book Number-13: 978-1-926895-56-7 (Hardcover)

Library of Congress Control Number: 2013950233

Library and Archives Canada Cataloguing in Publication

Aliyev, Azad, author
Economic-mathematical methods and models under uncertainty/Azad Aliyev, PhD.

Includes bibliographical references and index.
ISBN 978-1-926895-56-7
1. Economics--Mathematical models. 2. Economics--Mathematical models--Computer simulation. 3. Economics, Mathematical. I. Title.

HB135.A45 2013 330.01'51 C2013-906456-7

Apple Academic Press also publishes its books in a variety of electronic formats. Some content that appears in print may not be available in electronic format. For information about Apple Academic Press products, visit our website at **www.appleacademicpress.com** and the CRC Press website at **www.crcpress.com**

ABOUT THE AUTHOR

Azad Gabil Aliyev, PhD

Azad Gabil Aliyev, PhD, is Associate Professor in the Department of Economy and Management of Fuel Energy Complex Fields at the Azerbaijan State Oil Academy and is also Dean of the faculty of Economics, International Economical Relations and Management since 2011. He is the author of over 84 publications, including five monographs and 76 scientific articles.

CONTENTS

LIST OF SYMBOLS

V = vector space

F = field

$\vec{x}$, $\vec{y}$, $\vec{z}$, $\vec{a}_1$, $\vec{a}_2$, $\vec{a}_3$, $\overrightarrow{BC}$, $\overrightarrow{BA}$ = vectors

R_n = finite-dimensional Euclidean space

λ_n and μ_n = arbitrary positive numbers

k_1 and k_2 = intersection points

t = time interval

E_3 = tangent of the slope

i and j = indices

η= contingency factor

α= contiguity angle

x = variable changes in time interval

$\mu_1 \geq 0$ and $\mu_2 \geq 0$ = arbitrary parameters

PREFACE

At present, the progress in the sphere of economics is impossible without using up-to-date apparatus of economic-mathematical simulation. A mathematical apparatus is used in any of the modern courses of economics. In this/other degree the graphs of different dependences are analyzed, mathematical processing of these or other statistical data are conducted. Necessity in development of means for maintaining the processes of analysis, estimation and making of social-economic decisions becomes obvious in accordance with confirmation of natural market economics. The up-to-date dynamically changing market economics as a complex system functioning at uncertainty conditions generates a number of problems that require adequate analysis, estimation and choice of well-founded decision. The problems as estimation of risks and making up of securities portfolio, financial analysis, marketing and management of economic processes are among them.

The central problem of economics is the problem of rational choice. In the planned economics (at any rate in micro level, i.e., on the level of separate enterprise) there is no choice, this means that the role of mathematical approach was very belittled. In the market economics conditions, when each production unit should independently make a decision, i.e., to make a choice, mathematical calculation becomes necessary. Therefore, the role of mathematical methods in economics steadily increases. Therewith, for achieving the greatest effect by modeling the market economics objects it is appropriate to use intellectual technologies allowing to realize economic analysis, prediction and planning at uncertainty conditions, that in addition to quality economic indices. This allows to take into the account of weakly formalized quality factors and interrelations.

The section "economic-mathematical methods and models" combines a complex of productional, economic and mathematical principles that are intended to represent the picture of the studied economic process. However, it should be noted that because of its complexity, there is no real synthesis of economic theory, statistics, mathematics and informatics to the present day. Such a situation is connected with a number of principal profound problems in the character of economic processes themselves that have not been studied enough. While trying to formalize economic situation, one can get a very complicated mathematical problem. For simplifying, it is necessary to introduce new assumptions frequently not proved from the economical point of view. Therefore, the researchers are incited to danger to be engaged in mathematical technique instead of analysis of true economic situation. The main and essentially, a unique means of struggle against this, is verification of conclusions of mathematical theory by experimental data.

At the present stage, the given section of economics science develops in three principal directions: creation of analytic models, statistical models, and also development of uncertainty conditions in economic processes. Each of these directions has its advantage, and shortages.

We should note the peculiarities of these directions. Development of economic-mathematical simulation is on the way of improvement of calculation methods and extension of the circle of applied problems of economics, on the other hand, is on the way of studying such new phenomena that form a new view to the nature of interrelations in economics science, that allows to embrace a wider circle of applied problems from a united point of view. This second way becomes more urgent in solving the problems of reliable planning and control of economic processes.

The suggested monograph was written in the light of the second way of development of theory.

The book consists of introduction and five chapters. The directions of development of the section "economic-mathematical methods and models" that in its turn embraces the construction methods of appropriate economic-mathematical models, mathematical methods for solving economical problems, theory of prediction of economic processes and their control including at uncertainty conditions with wide description of unaccounted factors influence on behavior of the given processes are stated in introduction.

Brief information's on theory of finite-dimensional vector space widely known in mathematics, and its applications to economic problems are given in Chapter 1. In this chapter, the author formulated some necessary theorems connected with the method of construction of a conjugated vector in finite-dimensional Euclidean space. Necessary and sufficient conditions allowing to conduct the operation of division of a vector by a vector in Euclidean space were suggested by means of the given conjugated vector. A series of examples are cited for approving the given method.

Chapter 2 is devoted to development of piecewise-linear economic-mathematical models with regard to influence unaccounted factors that were fixed on previous small volumes of finite-dimensional vector space.

Here, it is proved that representation of economic processes in finite-dimensional vector space in the form of mathematical models is connected with complexity of complete account of such important problems as spatial inhomogeneity of occurring economical process, incomplete macro, micro and social political information and also time changeability of multi-factor economic indices, their duration and change velocity. The above-enumerated one reduces the solution of the given problem to creation of very complicated economic-mathematical models of nonlinear form.

In this connection, in Chapter 2 we first suggest the postulate: "Spatial-time certainty of economic process at uncertainty conditions in finite-dimensional vector space;" and give the principle: "Piecewise-linear homogeneity of economic process at uncertainty conditions in finite-dimensional vector space;" and also suggest a science-based method of multi-variant prediction of economic process and its control at uncertainty conditions in finite-dimensional vector space.

These main points allowed firstly, to set up mathematical dependence of any n-th piecewise linear vector equation on the first piecewise-linear function and all spatial form influence functions of unaccounted parameters influencing on all preceding piece-wise linear intervals of economic event. Secondly, to create conditions for formalization of the criterion of mathematical representation of predicting vector function of economic event depending on the desired influence function of unaccounted parameters that were previously fixed in small volumes of finite-dimensional vector space. A hyperconic surface whose points of directrix will create range of expected values of economic vent in the further step of small volume of finite-dimensional vector space was obtained in the result of such a representation.

A number of practically important piecewise-linear economic-mathematical models with regard to influence of unaccounted factors in three-dimensional vector space was suggested in Chapter 3. In this chapter two and three-component piecewise linear models and their appropriate functions of prediction of economic process and their control at uncertainty conditions in three-dimensional vector space were worked out and appropriate numerical calculations on their construction were given.

In Chapter 4, the suggested theory is applied to economic-mathematical problems on a plane. Here, geometrical interpretation of the introduced unaccounted parameter λ_m and of the influence function of unaccounted factors $w_n(t,\lambda_m)$ is given in a coordinate variant, and also the numerical calculation method is suggested for economical problems with regard to of influence unaccounted factors on a plane.

A series of model problems were considered on a plane on which the construction method of piecewise-linear economic-mathematical functions were approved, and also the method of definition of prediction and control functions of economic event with regard to of influence unaccounted factors is given.

This book is the second, complemented edition of earlier published monograph "Economic-mathematical methods and models at uncertainty conditions in finite-dimensional vector space." The essential novelty and distinctive peculiarity of the given monograph from the previous version is represented in Chapter 5. In Chapter 5, for the first time by means of theoretical and calculation instrumentation stated in the previous chapters, perspective software is developed for computer simulation by constructing and multi-variant prediction of economic state by means of two-component piecewise-linear economic-mathematical models with regard to of influence uncertainty factors in four-dimensional vector space.

In conclusion, the author expresses his deep gratitude to all persons that suggested their remarks and arguments in preparation of the copyright to be published. The author hopes that the monograph will be a useful manual for teachers, research associates in the field of economics, economic-mathematical methods, management, finances and credit, banking, and also for post-graduate students of appropriate specialties of higher educational institutions.

All critical remarks on the content and form of the statement of separate chapters and sections will be accepted with thanks.

— A. G. Aliyev, PhD

INTRODUCTION

Economic-mathematical methods and models are one of the important sections of economics science. This section reveals the nature of secrets of economics both in production and social-economic systems, exposes quality and quantity interactions and regularities, and creates reliable planning means, prediction and control of economic processes.

The section "economic-mathematical methods and models" combines a complex of production, economic and mathematical disciplines that are intended for giving a general picture of the studied economic process. However, it should be noted that because of its complexity, there is no real synthesis of economic theory, statistics, mathematics and information science today. Such a situation is connected with a lot of principal profound problems in the character of the economic processes themselves that have not been studied enough.

Today, modeling is a unique means of formalization and conduction of investigations of economic processes. In the general form, we can determine a model as a conditional image (simplified representation) of a real object (process) that is created for more or less profound study. The investigation method based on development and investigation of models is said to be modeling. Necessity of modeling is conditioned with complexity and sometimes with impossibility of direct study of a real object (process). It is considerably accessible to create and study the preimages of real objects (processes), i.e., the models. We can say that theoretical knowledge on something, as a rule, is the totality of different models. These models reflect essential properties of a real object (process), though in fact the reality is more significant and richer.

Therefore, today this section of economics science in principle develops in three main directions of economic processes modeling: creation of analytic models, creation of statistical models, and development of theory of uncertainty in economic processes. Each of these directions has its advantages and shortages.

Analytic and Statistical Models

So, analytic models in economics are more rough take into account a few number of deficiencies, always require some assumptions and simplifications. On the other hand, the results of calculations carried out by them are easily visible; they distinctly reflect principal regularities inherent to the phenomenon. And the chief, analytic models are more adjusted for searching optimal decisions, they can give qualitative and quantitative representation of the investigated economic event, are capable to predict the event, and also to suggest appropriate method on control of similar economic processes.

But statistical models, compared with analytic models are more precise and detailed, don't require such rough assumptions, allow to take into account a greater number (in theory infinitely greater) of factors. But statistical models also have some shortages including awkwardness, bad visibility of the economic event, great rates of machine time, and the chief, extreme difficulty of search of optimal decisions that have to be found "to the touch" by guess and trial [1–11].

Thus, it is considerably difficult to comprehend the results of statistic modeling than the calculations on analytic models and appropriately, it is more difficult to optimize the solution (it must be "found by feeling" blindly). Tame combination of analytic and statistical methods in investigation of operations is the matter of art, feeling and experience of the researcher. Quite often, it succeeds to describe some "subsystems" singled out in a big system by analytic methods and then from this models as from "small bricks" to construct a building of a big, complex model.

For today, the best scientific results in the field of investigation of economic processes and operations are based as a rule on joint application of analytic and statistical models. And the analytic method enables in general lines, to investigate the phenomenon, outline the contour of basic regularities. Any specifications may be obtained by means of statistical models.

Simultaneously, note that from time to time, human will may interfere with the control of economic process. The type of economic process simulation in the course of which a human being may take part is called an "imitational modeling." So a man that is at the head of operation, depending on the developed situation may make this or other decision as a chess-player that looking at the chessboard chooses his next move. Then a mathematical model is put into operation that shows what change of situation is expected to this decision and to which consequences it will lead in a lapse of time. The next "current" decision is made already with regard to a real new situation and etc. As a result of multiple repetition of such a procedure, the head as if "accumulates experience," learns in his own and others mistakes and gradually learns to make if not optimal, but almost optimal tame decisions.

The arising difficulties by constructing mathematical models of a complex economic system are the followings:

- if a model has many relations between the elements, if there are various nonlinear restrictions, and also if there is a great number of parameters and so on;
- frequently, real systems are subjected to different factors whose account by analytic way gives arise to great difficulties that are insurmountable because of their great number;
- comparison of the constructed analytic model and original under such approach is possible only near to the beginning of economic event. For from the beginning of the event, the degree of error between them significantly increases.

In principle, these difficulties stipulate application of a so-called "imitative" modeling [12].

One of the varieties of "imitative" modeling is the Monte-Carlo method known well in mathematics. The Monte-Carlo method is the numerical solution method of mathematical problems by means of random variables [13].

The Monte-Carlo method dates from 1949, after appearance of the paper "the Monte-Carlo method" It is considered that the creators of this method are the American mathematicians J. Neuman and S. Ulam. In the USSR, the first papers on the Monte-Carlo method were published in 1955–1956 years.

Oddly enough, the theoretical basis was known long ago. Furthermore, some problems of statistics were calculated sometimes with the help of random selections, i.e., in fact, by the Monte-Carlo method. However, before appearance of electronic computer machines (ECM) this method could not find a wide application, because to model random variables by hand is time-taking work. Thus, rise of the Monte-Carlo method as a very universal numerical method became possible only owing to appearance of ECM.

The name "Monte-Carlo" originates from the Monte-Carlo city in the principality Monaco that is famous with its game house.

The idea of the method is very simple and is in the following.

Instead of to describe the process by means of analytic apparatus (differential or algebraic equations), the "drawing" of a random event is conducted with the help of specially organized procedure including randomness and giving a random result. In reality, concrete realization of a random process each time develops quite differently; and also as a result of statistical modeling each time we get a new, different realization of the process under consideration. What can it give us? In itself nothing, as one case of treatment of a patient with the help of any medicine. It is another matter, if such realizations are a lot of. We can use this set of realizations as some artificially obtained synthetic material that may be processed by ordinary methods of mathematical statistics. After such a processing we can get any characteristics that we are interest in: probability of events, mathematical expectation variance of random variables and etc. By modeling random phenomena by the Monte-Carlo method, we use the randomness itself as an investigation device and we force it "to work for us" Frequently, such a way turns to be simpler than to attempt to construct an analytical model. For complex economic operations with a great number of elements (machines, peoples, organizations, additional means) where random factors complicatedly interlaced, the statistical modeling method as a rule becomes simpler than analytic one and very often becomes the only possible one.

As a matter of fact, any probability problem may be solved by the Monte-Carlo method, but it becomes justified if when the "drawing process" is simpler but not difficult than analytic calculation.

As a graphic example illustrating the imitation method we can cite the modeling of reproduction processes in oil-gas industry.

Contemporary stage of oil and gas industry development is characterized by the complication of relations and interactions of natural, economic, organizational, ecological and other factors of production both on the level of separate enterprises and oilgas recovering regions and field-wide level. In oilgas industry, the production differs by long periods, echelonment of production-technological process in time (search and prospecting and construction, oil, gas and condensate recovery), availability of log displacements and retardations, dynamic character of the used resources and other factors the values of many of which are of probability character.

The values of these factors systematically change owing to introduction into exploitation new deposits and also disconfirmation of the expected results on development. This compels the enterprises of oilgas industry to revise again the plans of reproduction of the basic funds and redistribute the resources in order to optimize the results of production-economic activity. The use of mathematical modeling methods including imitative ones may render essential help to persons preparing the project of economic decision. The essence of these methods is in repeated reproduction of the variants of planned decisions with subsequent analysis and choice of the most rational of these by the established system of criteria. A unit structural scheme integrating the functional control elements (strategically, tactical and operative planning) on the basic production processes of the field (search, prospecting, development, extraction, transport, oil-gas processing) may be created with the help of the imitative model. A wide application of the imitative method of modeling of reproduction processes in oil-gas industry may be found in the works [14–19].

Economic Cybernetics

With appearance of electron calculation machines (ECM) "the imitative method" found a generalized calculative character. It gave stimulus for creating a so-called new applied mathematical direction-"Economic cybernetics and mathematical methods in economics"

Necessity of systematic development of the fields of economics and enterprises, and also necessity of provision of different relations in the lines of labour and material resources, energetic economics, information and finances give to the total system a very dynamic and complicated character [1, 2, 6–8, 12, 20–32].

Under these conditions, the problems of optimal and proportional development of economics at present day are solved by means of economic cybernetics.

The field of economic cybernetics develops and grounds the notion, and also quality and quantity investigation methods of the effect of the laws of optimal and proportional growth of the components of the system of goods-production. Economic cybernetics reveals interactions of micro and macro systems. And by the same token, it allows to perceive best and control total economic mechanism. Practical application of the economic cybernetics methods enables to determine quantitatively the control parameters, to analyze and investigate models of economic development, structure and stability of economic processes, to reveal new possibilities for increasing productivity of labor and efficiency of production.

For example, the industrial economics is a dynamical system of special complexity. It consists of a great number of parts (subsystems) with numerous relations expressed by the stream of working forces, materials, energy, information and fiscal means.

Regain of balance of economic growth of industrial production is provided by a unique conceptual approach to economic phenomena called a system approach of economic cybernetics. This system approach is the main method of economic cybernetics.

The notion of economic-cybernetics system is introduced in connection with creation of fundamental bases on establishment of optimal, balanced and proportional growth conditions. The principal side of the economic-cybernetics system is that as an important element it contains a consciously acting man that executes the control, decision-making and control functions. Subject to these functions the issue is to provide optimal growth of economics. And, therewith, both internal and external influences should be taken into account. The higher the level of control of micro- and macroeconomics, the higher is the degree of complexity of control, management and decision-making functions. In this situation, a man gives it a dynamic character, high speed and capability to be improved. Thus, the economic-cybernetics systems represent a great complex of coordinated elements inter-conditioned within a more complicated structure and generating a single whole. And all of them are hierarchically ordered and interstipulate each other.

By analyzing the functions, behavior and structure of economic-cybernetics system, their dynamical character formed under the influence both of objective regularities and conscious actions of a man should be taken into account. The goal of investigations of economic-cybernetic systems, the relation between them and specific processes occurring in these systems is to order these entire phenomenon at concrete spatial and time conditions. Therefore, the notion of "economic space" and "economic time" are the basic categories of economic cybernetics.

The notion "economic space" contains the generalized representation of the properties of economic-cybernetic system in the input space and its development. In this sense, a space of industrial production is understood as a space of its possible motions within the planned economic growth. The essential indication of the space of industrial production is its unity and indivisibility.

The category of "economic time" reflects the totality of economic processes, their duration and velocity by which they are realized in economic-cybernetics systems. It should be noted that there is a close connection between the notion of economic system, economic space and economic time. Spatial-time aspect of studying economic-cybernetic systems allows to reveal well the motion and structure of processes.

Uncertainty in Economic Processes

Recently there are a lot of scientific publications in the world periodic literature where a great attention is paid to the uncertainty problem in economic operations and processes [14, 17, 26, 31, 33–42].

Development of contemporary society is characterized by the increase of technical level, complication of organizational structure of production, intensification of social division of labor, high requirements to the methods of planning and economic management.

On one hand, the globalization of the contemporary world economics with its increasing number of participants dictates to provide rational interrelation of participants of the game with regard to influence factors of high mobility of external medium.

On the other hand, a great stream of the primary information makes this task almost impracticable.

In other words, because of great flow of information they defy qualitative program processing. Especially, this concerns some factors of uncertainty character and complex structurization of information and formalization of its refining processes.

Taking into account that any organization situates and operates in the medium, the action of all organizations without exception is possible only if the medium admits its realization. The internal medium of the organization is the source of its vital forces, an external medium is the source supplying the organization by the resources necessary for maintaining its internal potential on the proper level.

The contemporary external medium of enterprises is characterized by the highest of degree complexity, dynamism and uncertainty. Capability of the enterprise to adapt to alternations in external medium is an important condition of its vital activity. Furthermore, in all increasing number of cases this is the main condition of survival and development. Therefore, the participants of the system realizing the economic operation, on one hand should steadily become aware of new character of changes in the environment and react effectively. On the other hand, it is necessary to have in mind that the organizes-participants themselves generate changes in the external medium producing new for instance types of goods and services, using new types of raw materials, energy, equipment's, technologies therewith changing the character of the technological process itself.

In scientific literature there exists a lot points of views on the occasion of character of the structure of the external medium and its influence on economic process. But the approach according to which two levels: micro and micro medium are distinguished in the external medium of any organization, is the most extended one.

The problem of interrelation of organization and medium in science was considered in the papers of A. Bogdanov and L. fon Bertalanfi in the first half of the 20th century. However, in complex systems the values of the external medium for organizations was realized only at the sixties under conditions of amplification of crises phenomena in economics. This served as a starting point for intensive use of the system approach in theory and practice of control from whose position any organization was considered as an open system interacting with external medium. The further development of the given conception gave rise to situational approach according to which the choice of the control method depends on concrete situation characterized to a great extent by the certain external variables. The original external medium was considered as the given activity conditions uncontrolled by the manager. At present the opinion that in order to live and develop at contemporary conditions, the organization should not only to adapt to external medium by means of adaptation of its internal structure and behavior at the market, but also actively form external conditions of its activity steadily revealing threats and potential possibilities at external medium, is of priority. This situation is on the basis of strategic control used by the advanced firms at conditions of high uncertainty of external medium.

Usually, in economic operations, analysis of external medium is considered as an initial process of strategic control since it provides bases for defining both the missions and goals of the firm on the whole and for working out strategy of behavior allowing the firm to perform its mission and achieve its goals. It represents itself as a process by means of which the creators of strategic plan control the factor external with regard to the organization in order to determine possible threats for the firm. Analysis of environment helps to obtain important results. It gives organization time for prediction of possibilities, making plans for the case of unforeseen circumstances, development of the system of earlier prevention for the cases of possible threats, and development of strategies that may change the earlier threats to different kind favorable possibilities.

In order to estimate necessity of thorough analysis of environment of the organization, it is necessary to consider the characteristics of external medium that render direct influence on the complexity of its realization.

Firstly, interconnectedness of the external medium factors belongs to the given characteristics. Under it we understand the level of the force with which the change of one factor influences on other factors. The fact of interconnectedness becomes significant already not only for the markets of the country or region, but also for the world market.

The heads can't consider the external factors separately. Recently, the specialists introduced the notion of "chaotic changes" for describing the external medium that is characterized by the quicker rates of change and stronger interconnectedness.

Secondly, we can note such a characteristics as the complexity of external medium. This is the number of factors to which the organization should react, and also on the level of variation of each factor. If we speak about the number of external factors to which the enterprises react, then if state resolutions, repeated renewal of contacts with trade unions, some interested groups of influence, numerous competitors and rapid technological changes, we can affirm that this organization is in a more complicated environment than for instance an organization worried by the actions of just several dealers, several competitors in non-availability of trade unions and slow change of technology.

Thirdly, we must distinguish the mobility of the medium. Under this we understand the rate of change in the environment of the organization. Many investigators note the tendency that the environment of contemporary organizations changes with rapidly increasing rates. However, this dynamics is general. There are organizations around of which the environment is especially mobile.

For example, as a result of investigations, it was revealed that the rate of change of technology and parameters of competitive struggle in pharmaceutical, chemical and electronic industry is higher than in other fields. Quick changes happen in aviation, cosmic industry, production of computers, biotechnology and in the sphere of telecommunications. The less noticeable changes touch the construction, food industry, production of container and packaging.

Furthermore, the mobility of the environment may be higher for one subdivisions of the organization and lower for other ones. For instance, in a lot of firms, the department of investigations and development come across the higher mobility of environment, since it should follow all technological innovations. On the other hand, the production department may be immersed in relatively slowly changing medium characterized by stable motion of materials and labor resources. At the same time, if production powers are scattered about different countries of the world or about the regions of the country, or input resources are received from abroad, then the production process may be found at conditions of highly mobile external medium.

Fourthly, there exist one more characteristics of the environment as uncertainty. It is the function of quantity of information that organization or (a person) disposes on the occasion of a concrete factor, and also a confidence function in this information. If information is negligible and if there is a doubt in its accuracy, the medium becomes more uncertain then in situation when there is adequate information and it may be considered highly reliable. In globalization period, economic processes are considered with regard to influence of larger number of external factors, therefore a more and reliability considerably and more information is required. However, therewith the confidence on its accuracy and falls down. The more uncertain environment, the more difficult to make effective decisions in the field of prediction and control.

The above listed characteristics of environment say on high dynamism and variation character of changes occurring in it. This imposes on the head of the problem as precise as possible prediction, estimation and analysis of the developed environment of the enterprises (firms) in order to establish in advance the character and power of possible threats, and this allows to work out and correct the chosen strategy. Such kind of control assumes earlier revealing of "weak signals" of any changes both interior and exterior to the enterprise and quick reaction to them. Therewith, and event and phenomena occurring at the environment of the enterprise should be steadily observed (monitoring). Thus, the so-called environment and its steadily variability contributes an essential share of uncertainty of proceeding of this or other economic process.

The complexity of solutions of economic problems is also in unavailability of explicit and complete definition of the uncertainty notion in economics, in unavailability of its proper classification, and also in unarialability of reliable mathematical representation of the phenomenon "uncertainty"

All these mentioned ones, increase the probability of risk and bankruptcy, lowers the degree of economic efficiency and etc. In this connection, there arises an utter necessity in detailed investigation of varieties of uncertainty phenomena in economic processes and their typical peculiarities.

In addition to what has been mentioned, environment, and also different external and internal factors also influence on the functioning of the economic system. These factors are the followings:

- climatic and meteorological conditions;
- economic risk (market economics);
- incompleteness of the structural construction of the system;
- difficulties of formalization and structurization of economic problems;

- insufficiency and sometimes surplus of information;
- instability (including political) factor;
- other factors.

The above listed hardly taken into account factors are combined to one general summarizing factor and we call it one of the uncertainty factors.

Such a summarizing uncertainty factor possesses the following properties:

- the factor of weak structurization of the system;
- the factor of stochastic property of the environment;
- risk factor;
- insufficient information factor;
- instability factor.

For consideration of such conglomerate of uncertainty factors we use the mathematical statistics apparatus that on the base of data for a preceding period works out probability of realization of a random event. For example, we can distinguish five main types of stochastic property of environment:

- instability of social-economic processes;
- sharp change of meteorological conditions;
- epidemic, diseases of the personnel;
- earthquakes, floods and other natural disasters;
- social-economic shocks of the society and etc.

Character of spatial inhomogeneity and many-dimensionality of the occurring economic process, changeability and rate of time change of many factor economic indices repeatedly complicates the solution of the given problem.

Fuzzy-Set Analysis of Economic Processes at Conditions of Essential Information Uncertainty

For the last 20–30 years, a flow of publications on application of fuzzy sets in economic and financial analysis in abroad increases. International Association for Fuzzy-Set Management and Economy (SIGEF) regularly approves new results in the field of fuzzy-set economic investigations. To day, in the world publication there are hundreds of monographs. Now the International scientific school in the former post Soviet space that also includes the Azerbaijan Republic, is formed [24, 26, 41, 43–54].

Note that theory of probability as a mathematical tool for modeling economic operations, and processes has taken its roots in economic analysis comparatively long ago (more than half a century ago). Fuzzy sets are a new, rather unhabitual mathematical tool in economic investigations. The founder of this direction is the world-known scientist Loutfi-Zadeh and his school [26, 52, 55].

So, beginning with the end of eighties, fuzzy-set applications to economic investigations began to be isolated from the general mathematical theory of fuzzy set as in its time the direction of intellectual computer systems and the systems based of fuzzy knowledge's, were isolated. This happened because economics is not engineering but a specific object of scientific investigations including strategy and tactics of economic activity "developing within social-historical formation based around the shaped production forces and production relations that covers all the links of commodity, distribution, commodity circulation and consumption of material benefits" Therefore, mathematicians developing the methodical part of applied theory we compel to have additional classification of economists in order to understand in detail the processes occurring in economics and to be scientifically analyzed.

Successful application of probability methods in statistics of the late 19th century (by investigating wide-range and statistically homogeneous demographic processes) extended the methods of probability theory in all the spheres of life especially with development of engineering cybernetics in the second part of the 20th century. The use of probabilities in accounting randomness, uncertainty and expectedness of events took an exclusive character. The most proved application was found in the events of mass character, more exactly, in queering theory and in engineering theory of reliability.

However, beginning with fifties, in academic science there appeared some works doubting about total applicability of probability theory to consideration of uncertainty. The authors of these works naturally remarked that classic probability was axiomatically determined as characteristics of general totality of statistically homogeneous random events. In the case when statistical homogeneity is unavailable, the application of classic probabilities in analysis is illegal.

The fundamentals works of Savage, Poy, Kayberg, Fischbern, de Finnetti and others where introduction of non-classic probabilities having no frequency sense but expressing cognitive activity of the investigator of random processes or of a person forced to make decisions in no information conditions were grounded, become the reaction to these entirely grounded remarks. The subjective (axiological) probabilities appeared in such a way. Therewith, a great majority of scientific results from classic theory of probability moved to theory of axiological probabilities, and in particular, logical-probability schemes of complex events based on the sorting of deductive conclusion of integral probabilities of complete set of initial conjectures on realization of simple events contained with the constituents in the complex event under consideration. These schemes were called implicative ones.

However, appearance of nonclassic probabilities was not an only reaction to the arising problem. It is necessary to note also a splash of interest to minimax approaches and also generation of theory of fuzzy sets. The goal of the minimax approaches is to reject from the considerations of uncertainty "by the weight method" i.e., when some expected integral effect is estimated, its formula is not a convolution of single effects when expert estimations or probabilities of realization of these effects emerge in place of weights of such a convolution. From the field of admissible realizations (scenarios) minimax methods choose two of them under which the effect accepts sequentially maximal or minimal value. Therewith, the decision-making person (DMP) should react to the situation so that he could achieve the best results at the worst conditions. It is assumed that such a behavior of DMP is the most optimal.

Opposing the minimax methods, the investigators remark that the expectedness of the worst scenarios may be found very lower and to dispose the decision making system to the worst result means to make unjustified higher expenditures and create groundless levels of all possible reserves. The use of the Hurwitz method according to which two external scenarios (the worst and the best) are considered jointly, and a parameter λ whose level is given by DMP plays in place of the weight in the convolution of scenarios, is the compromise way of application of minimax approaches. The more λ, the more optimistic is DMP. The modified interval-probability method of Hurwitz takes into account additional information on relation of probabilities of scenarios with regard to the fact that the precise value of scenarios probabilities is unknown.

The second scientific direction is the theory of fuzzy sets founded in Loutfi-Zadeh's fundamental book [26].

The primary conception of this theory was to construct functional accordance between fuzzy linguistic descriptions (as "high",," "warm" and etc) and special functions expressing the belongings degree of the values of measured parameters (length, temperature, weight, etc.) to the mentioned fuzzy descriptions.

The so-called linguistic probabilities given not quantitatively but by means of fuzzy-semantic estimation are also given in [26].

Later on, the range of applicability of theory of fuzzy sets was essentially widened. Loutfi-Zadeh himself determined fuzzy sets as a tool for constructing theory of possibilities [26, 55, 56].

Ever since scientific categories of randomness and possibility, probability and expectedness obtain theoretical delimitation.

The next achievement of theory of fuzzy sets is the introduction of so called fuzzy numbers as fuzzy subsets of specialized form that correspond to statements as "the value of the variable equals *a*."

Beginning with their introduction, it became possible to predict the further values of the parameters that are expectedly change in the established rated range. A collection of operations is performed on fuzzy numbers that are reduced to algebraic operations with ordinary numbers by giving definite reliability interval (belonging level).

Applied results of theory of fuzzy set were not hard to wait. For example, today a foreign market of so called fuzzy controllers, variety of which is established even in washing machines of widely advertised mark LG has a capacity of milliard dollars. The fuzzy logics as a model of human thinking processes have been built into the system of artificial intellect and in automated means of decision-making maintenance (in particular, to the system of control of technological processes).

We should note the following. The essential advantage of theory probabilities is a centuries-old historical experience of use of probabilities and logical schemes on their base. But when the uncertainty with respect to the further state of the investigation object loses the features of statistical uncertainty and classic sense of homogeneity of the process, classic probability as a characteristics of mass processes measurable in the course of tests go to nonexistence.

Falling-off in information situation challenge to life subjective probabilities, however there arises a reliability problem of probability appreciations. Appropriating point wise value to probabilities in the course of some virtual bet, the DMP proceeds from the arguments of own economic or other preferences that may be distorted by misrepresented expectations and bents. This remark is valid also in the case when not the DMP but an incidental expert appreciates the possibility.

By choosing appreciations of subjective probabilities, the Gibbs-Janes known principle is cited: when among all probability distributions agreed with input information on uncertainty of the appropriate index, it is advisable to choose that one to which the greatest entropy responds. A lot of researchers resorted to this principle for grounding probability conjectures in the structure of assumptions of initial model. However, natural objection against this principle propounded recently, is that the principle of maximum of entropy doesn't automatically provide monotonicity of the criterion of the expected effect. Hence, it follows that the principle of maximum of entropy should be complemented with boundary conditions of applicability of this criterion by choosing probability distributions.

In the case of application of fuzzy numbers to prediction of parameters, the DMP should not shape point wise probability appreciations but give rated corridor of values of predictable parameters. Then the expert appreciates the expected effect just in the same way as a fuzzy number with its own rated scatter (fuzzy degree). Here there arises engineering advantage of the method based on fuzzyness, since the researcher operates not only with oblique appreciations (where we refer probability as well) but with direct project data on statter of parameters and this is a well known practice of interval approach to projected appreciations.

Concerning the appreciations of the decision making risk at uncertainty conditions, the subjective-probability and fuzzy-set methods represent the same possibilities to the researcher. The stability degree of decisions is verified in the course of analysis of sensitivity of decision to fluctuations of initial data and this stability may be analytically appreciated.

So, tradition stands on the side of probability methods, convenience in engineering application and high degree of well-foundedness stands on the side of fuzzy set approaches, since all possible scenarios of development of events (generally speaking, generating a continuous spectrum) get into fuzzy-set calculation. We can't say it about the Hurwitz scheme adjusted to a finite discrete set of scenarios. And then the exclusive of quality interpretation of quality factors expressed in the terms of natural language belongs to fuzzy sets.

Piecewise-Linear Economic-Mathematical Models at Uncertainty Conditions

For the last 15 years in periodic literature there has appeared a series of scientific publications that has laid the foundation of a new scientific direction on creation of piecewise-linear economic-mathematical models at uncertainty conditions in finite dimensional vector space (4, 14, 17, 33–36, 52, 57–67].

Representation of economic processes in finite-dimensional vector space, in particular in Euclidean space, at uncertainty conditions in the form of mathematical models in connected with complexity of complete account of such important issues as: spatial inhomogeneity of occurring economic processes, incomplete macro, micro and social-political information; time changeability of multifactor economic indices, their duration and their change rate.

The above-listed one in mathematical plan reduces the solution of the given problem to creation of very complicated economic-mathematical models of nonlinear type.

In this connection, it was established in these works that all possible economic processes considered with regard to uncertainty factor in finite-dimensional vector space should be explicitly determined in spatial-time aspect. Owing only to the stated principle of spatial-time certainty of economic process at uncertainty conditions in finite dimensional vector space it is possible to reveal systematically the dynamics and structure of the occurring process. In addition, imposing a series of softened additional conditions on the occurring economic process, it is possible to classify it in finite-dimensional vector space and also to suggest a new science-based method of multivariant prediction of economic process and its control in finite-dimensional vector space at uncertainty conditions, in particular, with regard to unaccounted factors influence.

In the suggested monograph the author states the postulate "Spatial-time certainty of economic process at uncertainty conditions in finite-dimensional vector space",," gives the principle of "piecewise-linear homogeneity of economic process at uncertainty conditions in finite-dimensional vector space"

On these fundamental bases the author developed a special theory on construction of piecewise-linear economic-mathematical models with regard to unaccounted factors influence in finite-dimensional vector space, and suggested a new approach to the solution of the problem of multivariant prediction of economic process and its control at uncertainty conditions (with regard to influence of unaccounted economic and social-political factors) in finite-dimensional vector space [14,17,33-36].

For simplifying the applicability of the present theory to a wider class of economic processes, for making it accessible to a wider circle of specialists, the author developed a software for computer modeling applied by constructing and multivariant prediction of economic state by means of n-component piecewise economic-mathematical models with regard to influence of unaccounted factors in m-dimensional vector space.

The given program was written in the Matlab language, was successfully approved on numerous examples and allows to obtain prediction data coinciding with the conclusions of the earlier developed theory.

Here a perfect correspondence with the earlier developed graphic representation of piecewise-linear economic-mathematical models with regard to unaccounted factors influence on a plane was obtained for the case of upwards and downwards convexity, and also in the issue of establishment of the range of the predictable function that certifies its reliability.

The given program is also successfully approved on the sinusoidal type models [14, 17, 33–36].

REFERENCES

1. Allen, R. Mathematical economy. Publ. house Inostrannaya literature, Moscow, **1963**.
2. Ashmanov, S. A. Introduction to mathematical economy. Nauka, Moscow, **1984**.
3. Bagrinovskii, K. A.; Matyushok, V. M. Economical-mathematical methods and models.; Publ. house "Nauka," Moscow, **1999**.
4. Bellman, R.; Dynamical programming.; Publ. house inostr. Literature, Moscow, **1960**.

5. Braverman, E. M. Mathematical models of planning and control. Ed. M. "Nauka," Moscow, **1976,** 366.
6. Dantsing, J. Linear programming, its generalizations and applications, Moscow, **1966**.
7. Dolan, E. J.; Lindsay, D. Microeconomics, Russian translation, Moscow, **1994**.
8. Kantarovich, L. V.; Gorstko, A. B. *Optimal decisions in economics.* Ed. M. : "Nauka," Moscow, **1972,** 231.
9. Konyukhovskii, P. Mathematical investigations of operations in economics, Publ. house Piter, St. Petersburg. **2000**.
10. Lourie, A. L. On mathematical solution methods of optimum problems by planning national social economy. "Nauka," Moscow, **1964,** 235.
11. Makarov, V. L.; Rubinov, A. M.; Levin, M. I. Mathematical models of economic interaction. Ed. M. "Nauka," Moscow, **1993**.
12. Bagrinovskiy, K. A. *Imitative modelling of transitional economi,* In: "Control of economics of transitional period" Issue 2 M.; "Nauka, Fizmatlit," Moscow, **1998**.
13. Sobol, I. M. *Monte-Carlo method,* Ed. M. "Nauka," Moscow, **1985**.
14. Aliyev Azad, G.; Ecer, F. Tam olmayan bilqiler durumunda iktisadi matematik metodlar ve modeler, Ed. NUI of Turkiye, ISBN 975-8062-1802, Nigde. **2004,** 223.
15. Aliyev Azad, G. A model for defining the required profit and price for accumulation of the planned volume of proper financial means in fuzzy condition. Journal "Knowledge" "Education" Society of Azerbaijan Republic, Social Sciences 3–4, 1683–7649, Baku, **2005,** 6.
16. Aliyev Azad, G. Persepectives of development of the methods of modelling and prediction of economic processes. Izv. NAS of Azerbaijan, Ser. of hummanitarian and social sciences (economics), Baku, **2006,** *2,* 8.
17. Aliyev Azad, G. Development of dynamical model for economic process and its application in the field of industry of Azerbaijan. Thesis for PhD, Baku, **2001,** 167.
18. Aliyev Azad, G.; Farzaliyev, M. M. Some methodological aspects of developments of economic-mathematical models. ASOA Reports of Republishing Scientific-practical conference "Development of Sumgait in contemporary market conditions, problems and perspectives. SSU. Sumgait, **2007,** 2.
19. Belkin, V. D. A unit level prices and economic measuring on their base. Ed. M. "Economizdat," Moscow, **1963,** 321.
20. Vedutka, N. I. Economic cybernetics. Ed. M. "Nauka and Teknic," Moscow, **1971,** 318 .
21. Bugrov Ya. S.; Nikolskiy, S. M. *Elements of linear algebra and analytical geometry.* Ed. M. "Nauka," Moscow, **1980,** 175.
22. Wiener, N. *Creator and robot.* – Ed. M. "Progress," Moscow, 1966.
23. Gass, S. *Linear programming,* Ed. M. "Fizmatlit," Moscow, **1961,** 303.
24. Gunin, G. A. *Factors of practical application of artificial neural networks to prediction of financial time series.* In: Economic cybernetics: system analysis in economics and control, "Nauka," Moscow, 2001.
25. Dobrovolskiy, V. K. Economic-mathematical models. Ed. "Naukovo Dumka," Kiev. **1975,** 280.
26. Zadeh, L. Notion of linguistic variable and its application to making approximate decisions. Ed. M. "Mir," Moscow, 1976.

27. Imanov, K. D. Models of economic prediction. Ed. NAS of Azerbayjan "Elm," Baku. 1988.
28. Kantorovich, A. V.; Krylov, V. I. Approximate methods of higher analysis. Ed. M. "Fizmatlit," Moscow, **1962,** 590.
29. Kats, A. I. *Dynamic economic optimum (General criterion)*. Ed. M. "Ekanomika," Moscow, **1970,** 200.
30. Menesku, M. Economic cybernetics. Ed. M. "Ekonomic," Moscow, **1986,** 230.
31. Fedorenko, N. Issues of economic theory. Ed. M. "Nauka," Moscow, **1994,** 204.
32. Halmosh, P. R. Finite-dimensional vector spaces. Moscow, Ed. M. "Fizmatlit," Moscow, 1963.
33. Aliyev Azad, G. Economic-mathematical methods and models with regard to incomplete information. Ed. "Elm," ISBN 5-8066-1487-5, Baku. **2002,** 288.
34. Aliyev Azad, G. Economic-mathematical methods and models in uncertainty conditions in finite-dimensional vector space. Ed. NAS of Azerbaijan "Information technologies." ISBN 978-9952-434-10-1, Baku. **2009,** 220.
35. Aliyev Azad, G.; Theoretical bases of economic-mathematical simulation at uncertainty conditions, National Academy of Sciences of Azerbaijan, "Information Technologies," ISBN 995221056-9, Baku. **2011,** 338.
36. Aliyev Azad, G. Economical and mathematical models subject to complete information. Ed. Lap Lambert Akademic Publishing. ISBN-978-3-659-30998-4, Berlin. **2013,** 316.
37. Abbasov A. M.; Kuliyev, T. A. Economics in uncertainty. Ed. Azerbaijan. State Economics University. "Skazka," Baku. **2002,** 110.
38. Lomov, S. A. Power boundary layer in problems with small parameter. Dokl. AN SSSR, Moscow, **1963,** *148(3),* 5.
39. Neumman, J.; Margenstern, O. Theory of games and economic behavior, Ed. M. "Nauka," Moscow, 1970.
40. Trukhayev, R. I. Decision making models in uncertainty conditions. Ed. M. "Nauka," Moscow, 1981.
41. Aliyev, R. A.; Aliyev, R. R. Soft computing. Ed. "Cashioglu." Baku. **2004,** 620.
42. Imanov, K. D. Problems of economic uncertainty and Fuzzy models. Ed. "Elm," Baku. **2010,** 322.
43. Aliyev, R. A, Bijan Fazlollahi, Aliev, R. R. Soft Computing and its Application in Business and Economics. Ed. Spring Berlin Heidelberg, ISSN 3-540-22138-7, New York. **2004,** 450.
44. Aliyev, R. A, Bonfig K, Aliew, F. Soft Computin.; Ed. "Verlag Technic," Berlin. **2000,** 280.
45. Buckley, J. Solving fuzzy equations in economics and Finance. Fuzzy Sets and Systems, # 48, USA. 1992.
46. Davis, J. E-commerce still a risky business. E-commerce Times. USA. 1999.
47. J. Gil-Aluja. Elements for theory of decision in uncertainty. Ed. Kluwer academic Publishers. Ainty theory in Enterprises management. Ed. Springer. **1999**.
48. Jorion, P. Value-at-Risk: The New Benchmark for Managing Financial Risks. McGraw Hill Trade, ISBN 0071355022. **2000**.

49. Kaufmann, A.; Gupta, M. Introduction to Fuzzy Arithmetic: Theory and Applications. Van Nostrand Reinhold, ASIN 0442008996. **1991**.
50. Peray, K. Investing in mutual funds using fuzzy logic. St. Lucie Press, USA. **1999**.
51. Aliyev Azad, G. Search of decisions in fuzzy conditions. Izv. Visshikh tekhnicheskikh uchebnikh zavedeniy Azerbayjan. ASOA, ISSN 1609-1620, Baku. **2006,** *5(45),* 6.
52. Belman, R.; Zadeh, L. Decision makings in fuzzy conditions, In book: Issues of analysis and decision making procedure, Ed. M. "Mir," Moscow, 1976.
53. Coffmann, A.; Aluha, H. H. Introduction of theory of fuzzy sets in control of enterprises. Ed. "Vysshaya shkola," Minsk. 1992.
54. Ryzhov, A. P. Elements of theory of fuzzy sets and measuring of fuzzyness. Ed. M. "Dialog" MSU, Moscow, 1998.
55. Zadeh, L. A. Fuzzy sets as a basis for a theory of possibility. Fuzzy Sets and Systems, **1978**, *1,* 1.
56. Zadeh, L. A. Soft Computing and Fazzy Logic, IEEE Software **1994,** *11(6),* 11.
57. Aliyev Azad, G. Development of software for computer modeling of two-component piecewise-linear economic-mathematical model with regard to uncertainty factors influence in a four-dimensional vector space. Vestnik KhGAP, Khabarovsk. **2010,** *2(47),* 16.
58. Aliyev Azad, G. Prospecting's of software development for computer modeling of economic event by means of piecewise-linear economic-mathematical models with regard to uncertainty factors in three-dimensional vector space. Proc. of the III International scientific-practical conference "Youth and science: reality and future," V. Natural and applied problems. Nevinnomisk. **2010,** 6.
59. Aliyev Azad, G. Software development for computer modeling of two-component piecewise-linear economic-mathematical model with regard to uncertainty factors influence in *m*-dimensional vector space. "Natural and technical sciences" "Sputnik" publishing; ISSN 1684-2626, Moscow, **2010,** *2(46),* 12.
60. Aliyev Azad, G. Some aspects of computer modeling of economic events in uncertainty conditions in 3-dimensional vector space. Collection of papers of the XVIII International Conference Mathematics, Economics, Education of the VI International Symposium "Fourier series and their applications," Rostov-na-Donu. **2010,** 2.
61. Aliyev Azad, G. Software development for computer modeling of two-component piecewise economic-mathematical model with regard to uncertainty factors influence in 3-dimensional vector space. Publishing "Economics sciences," Moscow, **2010,** *3,* 8.
62. Aliyev Azad, G. Software development for multivariant prediction of economic event in uncertainty conditions on the base of two-component piecewise-linear economic-mathematical models in four-dimensional vector space. Publishing Vestnik of KhGAP, Khabarovsk. **2010,** *6(51),* 12.
63. Aliyev Azad, G. Software development for multivariant prediction of economic event in uncertainty conditions on the base of two-component piecewise-linear economic-mathematical model in m-dimensional vector space. Publishing "Economics sciences," Moscow, **2010,** *5 (66),* 10.
64. Aliyev Azad, G. Software development for multivariant prediction of economic event in uncertainty conditions on the base of two-dimensional piecewise-linear economic-mathematical model in three-dimensional vector space. Publishing "Economic, Statistic and Informatic" Vestnik of UMO, Moscow, **2010,** *3,* 10.

65. Aliyev Azad, G. Prospects of development of the software support for the computer simulation of the forecast of economic event using piece wise-linear economic-mathematical models in view of factors of uncertainty in m-dimensional vector spac. PCI, The third international conference "Problem of cybernetics and informatics," ISBN 978-9952-453-33-1, Baku. **2010,** *3,* 4.
66. Albegov, M. M. Short-term prediction in incomplete information conditions. Regional development and economic collaboration, Moscow, **1997**, *1.*
67. Leontyev, V. V. Economic essays. Ed. M. "Politizdat," Moscow, 1990.

CHAPTER 1

BRIEF INFORMATION ON FINITE-DIMENSIONAL VECTOR SPACE AND ITS APPLICATION IN ECONOMICS

CONTENTS

1.1 VECTOR SPACES AND THEIR PROPERTIES

1.1.1 FIELD

We preliminarily have given the definition to the number class, for example, to the class of real numbers or to the class of complex numbers. In order not to be restricted by any concrete class, we use the name of the numbers accepted in references, scalars. Below given are the general information on scalars which are divided into three groups of conditions: A, B, and C. [1].

(A) To each pair of scalars α, β there corresponds the scalar $\alpha+\beta$, called the sum of α and β, furthermore:

1. the addition is commutative, $\alpha+\beta=\beta+\alpha$;
2. the addition is associative, $\alpha+(\beta+\gamma)=(\alpha+\beta)+\gamma$;
3. there exists a uniquely determined scalar 0 (called a zero) such that $\alpha+0=\alpha$ for each scalar α;
4. To each scalar α there corresponds a uniquely determined scalar ($-\alpha$) such that $\alpha+(-\alpha)=0$;

(B) To each pair of scalar α, β there corresponds a scalar $\alpha\beta$ called the product of $\alpha\beta=\beta\alpha$, furthermore:

1. the multiplication is commutative, $\alpha\beta=\beta\alpha$;
2. the multiplication is associative, $\alpha\beta(\gamma)=(\alpha\beta)\gamma$;
3. there exists a uniquely determined nonzero scalar one (called a unit) such that for each scalar α;
4. to each nonzero scalar α there corresponds a uniquely defined scalar α^{-1} (or $\frac{1}{\alpha}$) such that $\alpha\alpha^{-1}=1$.

(C) The multiplication is distributive with respect to addition, $\alpha(\beta+\gamma)=\alpha\beta+\alpha\gamma$.

If the addition and multiplication were determined in some sets of objects (scalar) so that the conditions A, B, and C are fulfilled, then this set (with the above-mentioned operations) is called a *field*.

1.1.2 A VECTOR SPACE

In the following definition it is understood that some field of scalar F is given and all the scalars under consideration should be its elements.

Definition. The set V of elements called vectors and satisfying the following axioms is called a vector space:

(A) To each pair $\vec{x}, \vec{y}$ of the vectors from the set V there corresponds the vector $\vec{x}+\vec{y}$ called the sum of $\vec{x}$ and $\vec{y}$, furthermore:

1. the addition is commutative, $\vec{x}+\vec{y}=\vec{y}+\vec{x}$;
2. the addition is associative, $\vec{x}+(\vec{y}+\vec{z})=(\vec{x}+\vec{y})+\vec{z}$;
3. in V there exists a uniquely determined vector 0 (called origin) such that $\vec{x}+0=\vec{x}$ for each vector $\vec{x}$;
4. to each vector $\vec{x}$ from the set V there corresponds a uniquely determined vector $(-\vec{x})$ such that $\vec{x}+(-\vec{x})=0$.

(B) To each pair $\alpha, \vec{x}$, where α is a scalar, $\vec{x}$ is a vector from V, there corresponds the vector $\alpha\vec{x}$ called the product of α and $\vec{x}$ furthermore:

1. the multiplication by the scalar is associative, $\alpha\beta\vec{x}=(\alpha\beta)\vec{x}$, and
2. $1\vec{x}=\vec{x}$ for each vector $\vec{x}$.

(C)

1. Multiplication by the scalars is distributive with respect to addition of the vectors, $\alpha(\vec{x}+\vec{y})=\alpha\vec{x}+\alpha\vec{y}$, and
2. multiplication by the vectors is distributive with respect to addition of the scalars, $(\alpha+\beta)\vec{x}=\alpha\vec{x}+\beta\vec{x}$.

The relation between the vector space V and the main field F means that V is a vector space over the field F. If F is a field of real numbers, then V is called a real vector space; similarly, if the field F is a field of complex numbers, then V is called a complex vector space.

1.1.3 LINEAR DEPENDENCE OF VECTORS

Relations between the elements, which represents practical interest in the sequel are given below.

Definition. The finite set $\{\vec{x}_n\}$ of the vectors is said to be linear dependent if there exists an appropriate set $\{\alpha_n\}$ of scalars and not all of them equal zero, such that $\sum_n \alpha_n\vec{x}_n=0$. But if from $\sum_n \alpha_n\vec{x}_n=0$ it follows that $\alpha_n=0$ for each n, then the set $\{\vec{x}_n\}$ is said to be linear independent.

The vector $\vec{x} = \sum_n \alpha_n \vec{x}_n$, will mean that $\vec{x}$ is a linear combination of the vectors $\{\vec{x}_n\}$.

On the basis of what has been stated, we formulate a fundamental theorem on linear dependence of vectors.

1.1.4 THEOREM.

The set of nonzero vectors $\vec{x}_1, \vec{x}_2, \ldots\ldots, \vec{x}_n$ is linear dependent if some $\vec{x}_k$, $2 \le k \le n$ is a linear combination of preceding vectors.

Definition. The set X of linear independent vectors such that each vector of the space X is a linear combination of the elements from X is said to be a basis (or a coordinate system) of the vector space V. On the base of this definition we can say: any vector space V is finite-dimensional if it has a finite basis.

Definition. Two vector spaces U and V (over one and the same field) are isomorphic if we can establish one-tozczc1 correspondence between the vectors $\vec{x}$ from U and $\vec{y}$ from V (say $\vec{y} = T(\vec{x})$) so that

$$T(\alpha_1 \vec{x}_1 + \alpha_2 \vec{x}_2) = \alpha_1 T(\vec{x}_1) + \alpha_2 T(\vec{x}_2) \tag{1}$$

In other words, U and V are isomorphic if an isomorphism (similar to T) between them may be established.

Here under the isomorphism we understand one-tozczc1 correspondence preserving all linear relations. Recall that the objects that classic geometry of three-dimensional space studies, are the points of the considered space, straight lines, planes and etc. In this chapter, we study the analogies of these many-dimensional elements in general vector spaces. A nonempty subset m of the vector space V called a subspace or a linear manifold belongs to them. Thus, the notion of a subspace or a linear manifold is nothing but an analogy of school geometrical sense in the terms of vector space [1–7].

1.1.5 N-DIMENSIONAL REAL EUCLIDEAN SPACE

Based around the stated fundamental bases of a vector space, we have given the bases of n-dimensional real Euclidean space [1, 8].

The set of all possible systems $(x_1, x_2, ..., x_n)$ of real numbers is called n-dimensional real space and is denoted by R_n. We have denoted each system with one letter without an index.

$$\vec{x} = (x_1, x_2, ..., x_n) \tag{2}$$

and call a point or a vector of R_n. The numbers $x_1, x_2, ..., x_n$ are said to be the coordinates of the point (vector) or the components of the vector $\vec{x}$.

The two points,

$$\vec{x} = (x_1, x_2, ..., x_n),\ \vec{x}' = (x_1', x_2', ..., x_n') \tag{3}$$

are assumed to be equal if their appropriate coordinates are equal:

$$x_j = x_j' \ (j = 1, 2, ..., n) \tag{4}$$

In other cases, x_j and x_j' are distinct $(x_j \neq x_j')$.

The systems (vectors) $\vec{x} = (x_1, x_2, ..., x_n)$, $\vec{y} = (y_1, y_2, ..., y_n)$ are be put together, subtracted and multiplied by the numbers α, β, ... are real if R_n is a real space, and are complex if R_n is a complex space.

By definition, the sum of the vectors $\vec{x}$ and $\vec{y}$ is the vector,

$$\vec{x} + \vec{y} = (x_1 + y_1, x_2 + y_2, ..., x_n + y_n) \tag{5}$$

and the difference is the vector,

$$\vec{x} - \vec{y} = (x_1 - y_1, x_2 - y_2, ..., x_n - y_n) \tag{6}$$

The product of the number α and the vector $\vec{x}$ or of the vector $\vec{x}$ and the number α is the vector

$$\alpha\vec{x} = \vec{x}\alpha = (\alpha x_1, \alpha x_2, ..., \alpha x_n) \tag{7}$$

Obviously, the following properties are fulfilled:

(1) $\vec{x} + \vec{y} = \vec{y} + \vec{x}$,

(2) $(\vec{x} + \vec{y}) + \vec{z} = \vec{x} + (\vec{y} + \vec{z})$,

(3) $\vec{x} - \vec{y} = \vec{x} + (-1)\vec{y}$,

(4) $\alpha\vec{x} + \alpha\vec{y} = \alpha(\vec{x} + \vec{y})$,

(5) $\alpha\vec{x} + \beta\vec{x} = (\alpha + \beta)\vec{x}$,

(6) $\alpha\left(\beta\bar{x}\right) + \alpha\bar{y} = (\alpha\beta)\bar{x}$,

(7) $1 \cdot \vec{x} = \vec{x}$ (8)

where α, β are the numbers, and $\vec{x}, \vec{y} \in R_n$.

The number (nonnegative)

$$|\vec{x}| = \sqrt{\sum_{k=1}^{n} |x_k|^2} \tag{9}$$

is called the length or the norm of the vector $\vec{x} = (x_1, x_2, ..., x_n)$.

The distance between the points $\vec{x} = (x_1, x_2, ..., x_n)$ and $\vec{y} = (y_1, y_2, ..., y_n)$ is determined by the formula:

$$|\vec{x} - \vec{y}| = \sqrt{\sum_{k=1}^{n} |x_k - y_k|^2} \tag{10}$$

1.1.6 SCALAR PRODUCT

The scalar product of two vectors $\vec{x}=(x_1,x_2,....,x_n)$ and $\vec{y}=(y_1,y_2,....,y_n)$ of a real n-dimensional space is the number

$$(\vec{x},\vec{y})=\sum_{j=1}^{n}x_jy_j \tag{11}$$

Note that the space R_n where the scalar product is introduced is called Euclidean space.

The scalar product has the following properties:

(a) $(\vec{x},\vec{x})\geq 0$; therewith equality to zero holds if $\vec{x}=\vec{0}=(0,0,...0)$,

(b) $(\vec{x},\vec{y})=(\vec{y},\vec{x},)$

(c) $(\alpha\vec{x}+\beta\vec{y},\vec{z})=\alpha(\vec{x},\vec{z})+\beta(\vec{y},\vec{z})$ (12)

Indeed;

$$(\alpha\vec{x}+\beta\vec{y},\vec{z})=\sum_{j=1}^{n}(\alpha x_j+\beta y_j)z_j=\alpha\sum_{j=1}^{n}x_jz_j+\beta\sum_{j=1}^{n}y_jz_j=$$
$$=\alpha(\vec{x},\vec{z})+\beta(\vec{y},\vec{z})$$

Note that in real space in Eqs. (9) and (10) the sign of module (numbers) are replaced by the brackets:

$$|\vec{x}|=\sqrt{\sum_{j=1}^{n}x_j^2} \tag{13}$$

$$|\vec{x}-\vec{y}|=\sqrt{\sum_{j=1}^{n}(x_j-y_j)^2} \tag{14}$$

And the scalar product will be of the form:

$$(\vec{x},\vec{y})=\sqrt{\sum_{j=1}^{n}x_jy_j} \tag{15}$$

1.1.7 BUNYAKOVSKY INEQUALITY

From the properties A, B, and C of the scalar product it follows an important inequality called the Bunyakovsky inequality:

$$\left|\vec{x},\vec{y}\right| \le \sqrt{(\vec{x},\vec{x})}\cdot\sqrt{(\vec{y},\vec{y})} \tag{16}$$

Indeed, for a real number λ

$$0 \le (\vec{x}+\lambda\vec{y},\vec{x}+\lambda\vec{y}) = (\vec{x},\vec{x})+\lambda(\vec{y},\vec{x})+\lambda(\vec{x},\vec{y})+\lambda^2(\vec{y},\vec{y})=$$

$$=(\vec{x},\vec{x})+2\lambda(\vec{x},\vec{y})+\lambda^2(\vec{y},\vec{y})=a+2b\lambda+c\lambda^2,$$

where $a=(\vec{x},\vec{x}),\ b=(\vec{x},\vec{y}),\ c=(\vec{y},\vec{y})$.

Hence, it is seen that the square polynomial

$$a+2b\lambda+c\lambda^2 \ge 0 \ (-\infty<\lambda<\infty) \tag{17}$$

is nonnegative for any real λ. Therefore, its graph lies above the axis λ, and this may be only when the discriminant of the polynomial is nonnegative or equals zero, i.e. for $b^2-ac\le 0$ or $b^2\le ac$ and as a result, Bunyakovsky inequality, Eq. (16), is obtained.

In the terms of the components of the vectors $\vec{x}$ and $\vec{y}$, Bunyakovsky inequality, Eq. (16), is written in the form:

$$\left|\sum_{j}^{n} x_j y_j\right| \le \sqrt{\sum_{j=1}^{n} x_j^2}\cdot\sqrt{\sum_{j=1}^{n} y_j^2} \tag{18}$$

Thus, whatever were the real values x_j, y_j, inequality Eq. (18) is fulfilled.

By Eq. (9) we can write the Bunyakovsky inequality in the following way:

$$|(\vec{x}, \vec{y})| \leq |\vec{x}| \cdot |\vec{y}| \tag{19}$$

But then there exists and uniquely the number λ satisfying the inequality $-1 \leq \lambda \leq 1$ for which the following revise equality holds:

$$(\vec{x}, \vec{y}) = \lambda |\vec{x}| \cdot |\vec{y}| \tag{20}$$

Note that on the interval $[0, \pi]$ the Cost Function has a unique strongly decreasing inverse function with domain of values strongly decreasing inverse function with domain of values on $[-1,1]$. Therefore, for each λ $(-1 \leq \lambda \leq 1)$ there exists a unique angle φ $(0 \leq \varphi \leq \pi)$ such that $\lambda = \cos\varphi$. Thus, we proved the equality:

$$(\vec{x}, \vec{y}) = |\vec{x}| \cdot |\vec{y}| \cdot \cos\varphi \tag{21}$$

The number φ is said to be an angle between the n-dimensional vectors $\vec{x}$ and $\vec{y}$. But as matter of fact for $n > 3$ the vectors $\vec{x}$ and $\vec{y}$ are not real segments, they are mathematical abstractions.

The vectors $\vec{x}$ and $\vec{y}$ are said to be orthogonal if the scalar product equals zero:

$$(\vec{x}, \vec{y}) = \sum_{j=1}^{n} x_j y_j = 0 \tag{22}$$

From Eq. (21) it follows that in order the nonzero vector $\vec{x}$ and $\vec{y}$ be orthogonal, it is necessary and sufficient that the angle between them be $\varphi = \pi/2$.

1.1.8 MINKOVSKY INEQUALITY

Note an important property of the Minkovsky inequality:

$$|\vec{x} + \vec{y}| \leq |\vec{x}| + |\vec{y}| \tag{23}$$

or in the terms of components:

$$\sqrt{\sum_{j=1}^{n}(x_j + y_j)^2} \le \sqrt{\sum_{j=1}^{n} x_j^2} + \sqrt{\sum_{j=1}^{n} y_j^2} \tag{24}$$

We can prove it as follows:

$$|\vec{x} + \vec{y}|^2 = (\vec{x} + \vec{y}, \vec{x} + \vec{y}) = (\vec{x}, \vec{x}) + 2(\vec{x}, \vec{y}) + (\vec{y}, \vec{y})$$

Using Bunyakovsky inequality, Eq. (16), we have:

$$\begin{aligned} |\vec{x} + \vec{y}|^2 &\le (\vec{x}, \vec{x}) + 2\sqrt{(\vec{x}, \vec{x})} \cdot \sqrt{(\vec{y}, \vec{y})} + (\vec{y}, \vec{y}) = \\ &= (\sqrt{(\vec{x}, \vec{x})} + \sqrt{(\vec{y}, \vec{y})})^2 \end{aligned} \tag{25}$$

Hence, it follows Eq. (23), that yields the inequality,

$$\big||\vec{x}| - |\vec{y}|\big| \le |\vec{x} - \vec{y}| \tag{26}$$

because,

$$|\vec{x}| = |\vec{x} - \vec{y} + \vec{y}| \le |\vec{x} - \vec{y}| + |\vec{y}|$$

$$|\vec{y}| = |\vec{y} - \vec{x} + \vec{x}| \le |\vec{x} - \vec{y}| + |\vec{x}| \tag{27}$$

1.1.9 SEGMENT: PARTITION OF THE SEGMENT IN THE GIVEN RELATION

Give arbitrary points $\vec{x}, \vec{y} \in R_n$ and introduce the set of points (vectors):

$$\vec{z} = \lambda\vec{x} + \mu\vec{y} \quad (\lambda, \mu \ge 0, \ \lambda + \mu = 1) \tag{28}$$

determined by the nonnegative numbers λ, μ whose sum equals one. We have:

$$\vec{z} = (1 - \mu)\vec{x} + \mu\vec{y} = \vec{x} + \mu(\vec{y} - \vec{x}) \quad (0 \le \mu \le 1) \tag{29}$$

or

$$\vec{z} = \vec{y} + \lambda(\vec{x} - \vec{y}) \quad (0 \le \lambda \le 1) \tag{30}$$

It is seen from equality Eq. (29) that in three-dimensional space, the points $\vec{z}$ fill the segment connecting $\vec{x}$ and $\vec{y}$. The vector-radius $\vec{z}$ is the sum of the vector $\vec{x}$ and vector $\mu(\vec{y} - \vec{x})$, collinear with $(\vec{y} - \vec{x})$ (Fig. 1).

Thus, the set of points Eq. (28) is a segment $[\vec{x}, \vec{y}]$ in R_n connecting the points $\vec{x}$ and $\vec{y}$. For $\mu = 0$ $\vec{z} = \vec{x}$, and for $\lambda = 0$ $\vec{z} = \vec{y}$, for any $\lambda > 0$ $(\mu = 1 - \lambda > 0)$ $\vec{z}$ is an arbitrary point $[\vec{x}, \vec{y}]$.

By definition, the set of all points $\vec{z}$ of the form Eq. (28) is called the segment $[\vec{x}, \vec{y}]$ connecting the points $\vec{x}, \vec{y} \in R_n$. Here the following theorem is valid.

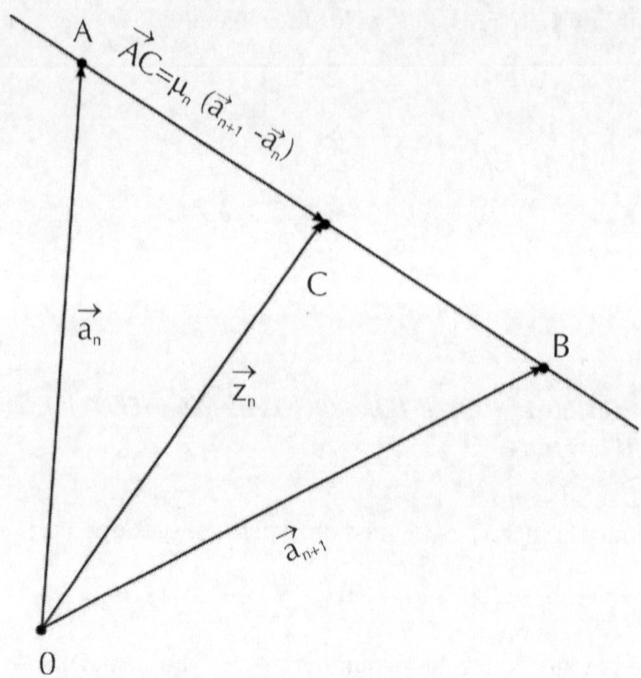

FIGURE 1 Dependence of the vector-point on the vectors $\vec{z}_n$ on the vectors $\vec{a}_{n+1}$.

1.1.10 THEOREM 1

The point

$$\vec{z} = \lambda\vec{x} + \mu\vec{y} \ (\lambda, \mu \geq 0 \,, \ \lambda + \mu = 1)$$

partitions the segment $[\vec{x}, \vec{y}]$ connecting the points $\vec{x}, \vec{y} \in R_n$ into the segments with the length being in the ratio $\mu : \lambda$.

Proof. From Eq. (30) it follows that $\vec{z} - \vec{x} = \mu(\vec{y} - \vec{x})$, and therefore the distance between the point $\vec{x}$ and $\vec{y}$ equals:

$$\left|\vec{z} - \vec{x}\right| = \mu\left|\vec{y} - \vec{x}\right| \tag{31}$$

From Eq. (30) where $\vec{z} - \vec{y} = \lambda(\vec{x} - \vec{y})$, the distance between the points $\vec{z}$ and $\vec{y}$ will equal:

$$\left|\vec{z} - \vec{y}\right| = \lambda\left|\vec{x} - \vec{y}\right| \tag{32}$$

From Eqs. (31) and (32) it follows:

$$\left|\vec{z} - \vec{x}\right| : \left|\vec{z} - \vec{y}\right| = \mu : \lambda$$

Q.E.D.

In conclusion we once more note that in three-dimensional space R_3 the points set:

$$\vec{z} = \lambda\vec{x} + \mu\vec{y} \,, \ \lambda + \mu = 1 \tag{33}$$

where λ and μ are of any sign, is a straight line passing through the points $\vec{x}$ and $\vec{y}$. We can see it from equality Eq. (30). But in space R_n, where $(n > 3)$ this set is said to be straight by definition.

1.2 EXAMPLES OF MANY-DIMENSIONAL PROBLEMS OF ECONOMICS

1.2.1 EXAMPLE 1

A group of students have gone off hiking expedition to some Europen capitals. To the end of the trip they revealed that they have 15 French franc, 10 Britain found sterling, 20 Netherlandish gulden and 25 German mark in their purse. The remainders composed the "Currency" vector:

$$\vec{a} = (15, 10, 20, 25)$$

The students decided to convent the currency into dollars and organize dinner. At the exchange point they learned the rate of exchange:

1 French franc = 1,000 dollars,
1 Britain pound sterling = 7,500 dollars,
1 Netherlandish gulden = 3,000 dollars,
1 German mark = 3,500 dollars.

Thus, there appears one more four-dimensional vector of the rate of exchange:

$$\vec{b} = (1000;7500;3000;3500)$$

In order to determine how many dollars for dinner, it is necessary to make the following calculation:

$$15 \cdot 1000 + 10 \cdot 7500 + 20 \cdot 3000 + 25 \cdot 3500 = 237500 \text{ dollars}$$

1.2.2 EXAMPLE 2

A commercial bank participating in construction of multi-storey parking at the center of Baku, undertook to take credits in three commercial banks: "Most-Bank", "Baku Business-Bank", "Capital Bank of Savings". Each of these banks gave credits in 20, 40 and 40 milliard dollars under the annual rate 40, 25 and 30%.

In the given example we speak about two vectors: a three-dimensional vector *f* credits $\vec{c} = (20; 40; 40)$ and a vector of rates $\vec{p} = (40; 25; 30)$.

Using a simple calculation, the manager of the commercial bank define how much they must pay in a year for the credits taken from the three banks:

$$20 \cdot 1,4 + 40 \cdot 1,25 + 40 \cdot 1,3 = 130 \; mlrd.dollar$$

In these examples, we can see rise of peculiar operation over the vectors from R_n, called a scalar multiplication of vectors.

1.2.3 EXAMPLE 3

Practically, it is very easy to discover collinearity of two vectors. The coordinates $\vec{a}_1, \vec{a}_2, ..., \vec{a}_n$ of the first vector $\vec{a}$ should be proportional to the coordinates $\vec{b}_1, \vec{b}_2, ..., \vec{b}_n$ of the second vector $\vec{b}$. A table of currency rate is a good example of collinear vectors. Each Friday a table of currency rates whose conditional fragment is given in Table 1, is published in financial appendix to the newspaper "Nedelniye novosti:"

TABLE 1 Table of conditional values of exchange of the currency rate.

	1$	1 DM	1 SF
A	1	2	3
1 $	1	0.6806	0.834
1 DM	1.4693	1	1.22541
1 SF	1.199	0.8161	1

Each column of Table 1 expresses the rate value of the unit of appropriate kind of currency. So the second column shows that for one DM one can get 68 cent or 82 centime.

Any two columns and any two lines of Table 1 are proportional, i.e., any vector-columns and any vector-lines are collinear, and in Table 1 it is easy to find the proportionality factor.

1.3 CONSTRUCTION OF A CONJUGATED VECTOR IN FINITE-DIMENSIONAL VECTOR SPACE AND ITS APPLICATION TO ECONOMICS PROBLEMS

In this section, we suggest a special method for constructing a conjugated vector in finite-dimensional Euclidean space, based on the suggested theorems. By means of the conjugated vector, necessary and sufficient conditions admitting to perform a symbolic operation of division of a vector by a vector in Euclidean space are suggested. In addition, some examples [9–11] for realization of the suggested method are given.

1.3.1 THEOREM 1

If in finite-dimensional Euclidean space R_n the two vectors $\vec{a}$ and $\vec{b}$ are collinear, in this case their unit vectors $\vec{a}_0$ and $\vec{b}_0$ will be conjugated to each other. Secondly, for the vector $\vec{a}$ of this space there will exist some modified vector $\vec{c}$ conjugated to it.

It follows from what has been said that by means of two collinear vectors $\vec{a}$ and $\vec{b}$ of Euclidean space we can construct some vector $\vec{c}$ conjugated to the vector $\vec{a}$, so that scalar product of their unit vectors $\vec{a}_0$ and $\vec{b}_0$ will equal unit, i.e.,

$$\vec{a}\cdot\vec{c}=1 \quad \vec{a}_0\cdot\vec{b}_0=1 \tag{34}$$

Introducing the modified vector $\vec{c}$ as a product of the vector $\vec{b}$ and some scalar $\eta=\frac{1}{|\vec{a}|\cdot|\vec{b}|}$, we can write Eq. (34) in the form:

$$\vec{a}\cdot\vec{c}=1 \tag{35}$$

where

$$\vec{c}\cdot\eta\cdot\vec{b}=\frac{\vec{b}}{|\vec{a}|\cdot|\vec{b}|} \tag{36}$$

and $\eta\cdot=\frac{1}{|\vec{a}|\cdot|\vec{b}|}$ is called a conjugation factor of the vectors $\vec{a}$ and $\vec{b}$.

1.3.2 PROOF

It is known that in finite-dimensional Euclidean space R_n, the scalar product of two collinear vectors equals the product of their module:

$$\vec{a}\cdot\vec{b}=|\vec{a}|\cdot|\vec{b}| \tag{37}$$

Taking into account

$$\vec{a}=|\vec{a}|\cdot\vec{a}_0,\ \vec{b}=|\vec{b}|\cdot\vec{b}_0 \tag{38}$$

where $\vec{a}_0$ and $\vec{b}_0$ are unit vectors, we represent Eq. (36) in the form:

$$\vec{a}_0\cdot\vec{b}_0=1 \tag{39}$$

It is seen from Eq. (39) that the scalar product of collinear unit vectors $\vec{a}_0$ and $\vec{b}_0$ equals one. Hence, it follows that the unit vector $\vec{b}_0$ is conjugated to the unit vector $\vec{a}_0$. Now, substitute expression Eq. (38) in Eq. (39), and get:

$$\vec{a}=\frac{1}{\frac{1}{|\vec{a}|\cdot|\vec{b}|}\cdot\vec{b}}=\frac{1}{\eta\cdot\vec{b}} \tag{40}$$

Introducing the denotation of the modified vector $\vec{c}$ as a product of the vector $\vec{b}$ and some scalar $\eta=\frac{1}{|\vec{a}|\cdot|\vec{b}|}$, we get:

$$\vec{a}\cdot\vec{c}=1 \tag{41}$$

Here, the scalar $\eta = \frac{1}{|\vec{a}| \cdot |\vec{b}|}$ is said to be a conjugation factor of the vectors $\vec{a}$ and $\vec{b}$. It is seen from Eq. (41) that for the vector $\vec{a}$ there exists some modified vector $\vec{c}$ conjugated to it, of the form:

$$\vec{c} = \eta \cdot \vec{b}$$

1.3.3 EXAMPLE 1

Let in three-dimensional vector space R_3 two collinear piecewise-linear vectors $\overrightarrow{BC} = \vec{a}_3 - \vec{a}_2 \in R_3$ and $\overrightarrow{AB} = \vec{a}_2 - \vec{a}_1 \in R_3$ (Fig. 2) be given.

The vectors (points) $\vec{a}_1$, $\vec{a}_2$ and $\vec{a}_3$ have the following form:

$$\vec{a}_1 = 2\vec{i} + 3\vec{j} + 4\vec{k}, \ \vec{a}_2 = 3\vec{i} + 5\vec{j} + 7\vec{k},$$

$$\vec{a}_3 = 5\vec{i} + 9\vec{j} + 13\vec{k} \tag{42}$$

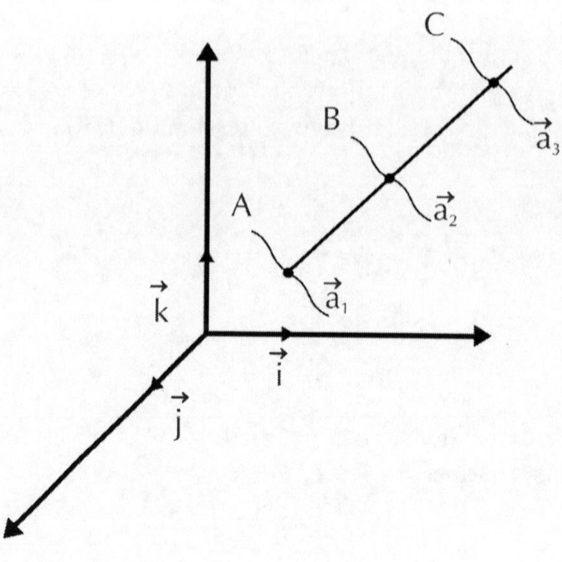

FIGURE 2 Dependence of the vector vectors $\vec{a}_1$, $\vec{a}_2$, $\vec{a}_3$.

Show the way for calculating symbolic operation of division of collinear vectors $\overrightarrow{BC}$ by $\overrightarrow{AB}$, i.e.,

$$\frac{\overrightarrow{BC}}{\overrightarrow{AB}}=\frac{\vec{a}_3-\vec{a}_2}{\vec{a}_2-\vec{a}_1} \tag{43}$$

According to theorem one, for realization of symbolic operation of division of one piecewise-linear vector by another vector, it is necessary to construct some piecewise-linear vector $\vec{c}$ conjugated to the vector of numerator (or denominator).

Taking into account collinearity of the vectors $\overrightarrow{BC}$ and $\overrightarrow{AB}$, by theorem one, the conjugated to the numerator vector $\overrightarrow{BC}$ will be the vector $\vec{c}$ equal.

$$\vec{c}=\eta\cdot\overrightarrow{AB} \tag{44}$$

Here η is a contingency factor between the collinear vectors $\overrightarrow{BC}$ and $\overrightarrow{AB}$ and equals:

$$\eta=\frac{1}{\left|\overrightarrow{BC}\right|\cdot\left|\overrightarrow{AB}\right|} \tag{45}$$

And the scalar product of the vectors $\overrightarrow{BC}$ by the conjugated vector $\vec{c}$ will be equal to 1, i.e.,

$$\overrightarrow{BC}\cdot\vec{c}=1 \tag{46}$$

Now, multiply and divide the numerator and denominator of vector fraction Eq. (43) by the modified vector $\vec{c}$ and taking into account conditions Eqs. (45) and (46), we get:

$$\frac{\overrightarrow{BC}}{\overrightarrow{AB}}=\frac{\vec{a}_3-\vec{a}_2}{\vec{a}_2-\vec{a}_1}=\frac{(\vec{a}_3-\vec{a}_2)\cdot\vec{c}}{(\vec{a}_2-\vec{a}_1)\cdot\vec{c}}=\frac{1}{\eta(\vec{a}_2-\vec{a}_1)^2}= \\ =\frac{\left|\vec{a}_3-\vec{a}_2\right|\left|\vec{a}_2-\vec{a}_1\right|}{(\vec{a}_2-\vec{a}_1)^2} \tag{47}$$

Substituting the value of the vectors $\vec{a}_1$, $\vec{a}_2$ and $\vec{a}_3$ by Eq. (42), define the number value of vector fraction Eq. (47) in the final form:

$$\frac{\overrightarrow{BC}}{\overrightarrow{AB}} = \frac{\vec{a}_3 - \vec{a}_2}{\vec{a}_2 - \vec{a}_1} = 2 \tag{48}$$

1.3.4 THEOREM 2

If in finite-dimensional Euclidean space R_n there is a scalar product between adjoint vectors $\vec{a}$ and $\vec{b}$, in this case for the vector $\vec{a}$ there will exist some modified vector $\vec{c}$ conjugated to it. And the formed vector $\vec{c}$ will equal the product of the vector $\vec{b}$ and some contingency factor η.

It follows from what has been said that for the vector $\vec{a}$ adjacent to the vector $\vec{b}$, for any contiguity angle $0 \le \alpha < 90°$ there will exist such a modified vector $\vec{c}$ conjugated to it that the scalar product of the vector $\vec{a}$ and $\vec{c}$ will be equal to unit, i.e.,

$$\vec{a} \cdot \vec{c} = 1 \text{ or } \vec{a} = \frac{1}{\vec{c}} \tag{49}$$

Here,

$$\vec{c} = \eta\vec{b} = \frac{1}{|\vec{a}| \cdot |\vec{b}| \cdot \cos\alpha}\vec{b} \tag{50}$$

and call $\eta = \dfrac{1}{|\vec{a}| \cdot |\vec{b}| \cdot \cos\alpha}$ the contingency factor between the adjacent vectors $\vec{a}$ and $\vec{b}$.

1.3.5 PROOF

Let in finite-dimensional Euclidean space R_n the scalar product of two adjoint vectors $\vec{a}$ and $\vec{b}$ (Fig. 3) be given:

$$\vec{a}\cdot\vec{b}=|\vec{a}|\cdot|\vec{b}|\cdot\cos\alpha \tag{51}$$

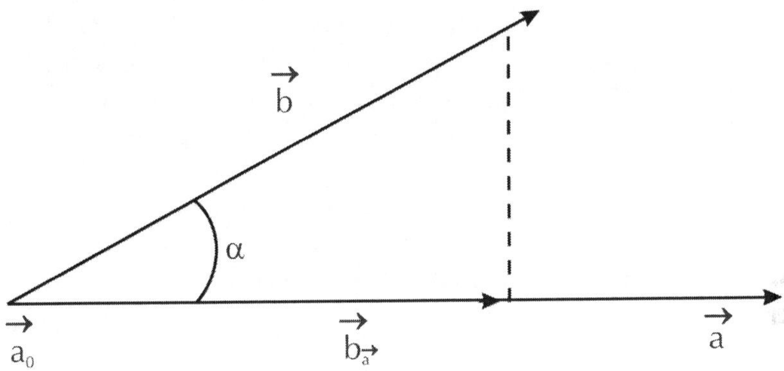

FIGURE 3 Interpretation of scalar product of two vectors.

Here

$$\vec{a}=|\vec{a}|\cdot\vec{a}_0,\ \vec{b}=|\vec{b}|\cdot\vec{b}_0 \tag{52}$$

Here $|\vec{b}|\cdot\cos\alpha$ is the projection of the length of the vector $\vec{b}$ on the direction of the vector $\vec{a}$; α is an contiguity angle between the vectors $\vec{a}$ and $\vec{b}$. Taking into account that the square of a unit vector $\vec{a}_0^2=1$, we represent Eq. (51) in the form:

$$\vec{a}\cdot\vec{b}=\vec{a}\cdot|\vec{b}|\cdot\cos\alpha\cdot\vec{a}_0 \tag{53}$$

Here the vector

$$\vec{b}_{\vec{a}}=|\vec{b}|\cdot\cos\alpha\cdot\vec{a}_0 \tag{54}$$

will be a vector-projection of the vector $\vec{b}$ on the direction of the vector $\vec{a}$ (Fig. 3).

In this case we can write Eq. (53) in the form:

$$\vec{a} \cdot \vec{b} = \vec{a} \cdot \vec{b}_{\vec{a}} \tag{55}$$

Hence, it follows that the scalar product of two adjoint vectors $\vec{a}$ and $\vec{b}$ is nothing but a scalar product of the vector $\vec{a}$ and the vector-projection $\vec{b}_{\vec{a}}$ of the second vector $\vec{b}$ on the direction of vector $\vec{a}$. Hence it follows that the vector $\vec{a}$ and vector-projection $\vec{b}_{\vec{a}}$ are collinear to each other. Therefore, from theorem one, their unit vectors will be conjugated to each other:

$$\vec{a}_0 \cdot \left(\vec{b}_{\vec{a}}\right)_0 = 1 \tag{56}$$

Here, the unit vector projection $\vec{a}_0 \cdot \vec{b}_0 = 1$ will equal:

$$\left(\vec{b}_{\vec{a}}\right)_0 = \frac{\vec{b}_{\vec{a}}}{\left|\vec{b}_{\vec{a}}\right|} \tag{57}$$

Now, taking into account Eq. (52), write Eq. (51) in the form:

$$\vec{a}_0 \cdot \vec{b}_0 = \cos\alpha \tag{58}$$

Hence it follows that for any $\alpha \neq 0$ the unit vectors $\vec{a}_0$ and $\vec{b}_0$ will not be conjugated to each other. Only in particular case, for $\alpha = 0$ these unit vectors, by their collinearity, will be conjugated to each other. Multiply the left and right sides of Eq. (58) by the scalar $\frac{1}{\cos\alpha}$ and introduce the denotation of the modified unit vector:

$$\vec{c}_0 = \frac{1}{\cos\alpha} \cdot \vec{b}_0 \tag{59}$$

We get:

$$\vec{a}_0 \cdot \vec{c}_0 = 1 \tag{60}$$

From Eq. (60) it follows that for any α changing on the interval $0 \leq \alpha < 90^\circ$, for the unit vector $\vec{a}_0$ there will exist some modified unit vector $\vec{c}_0$ conjugated to it. Now, substitute Eqs. (52) and (59) in Eq. (60) and we get:

$$\vec{a} = \frac{1}{\dfrac{1}{\left|\vec{a}\right| \cdot \left|\vec{b}\right| \cdot \cos\alpha} \cdot \vec{b}} = \frac{1}{\eta \vec{b}} \tag{61}$$

Introducing the denotation of the modified vector $\vec{c}$ as a product of the vector $\vec{b}$ and some scalar of the form $\eta = \dfrac{1}{\left|\vec{a}\right| \cdot \left|\vec{b}\right| \cdot \cos\alpha}$, Eq. (61) accepts the form:

$$\vec{a} \cdot \vec{c} = 1 \tag{62}$$

Here the constant $\eta = \dfrac{1}{\left|\vec{a}\right| \cdot \left|\vec{b}\right| \cdot \cos\alpha}$ is said to be a contingency factor between the adjacent vectors $\vec{a}$ and $\vec{b}$. From Eq. (62) it follows that for the vector $\vec{a}$ adjoint to the vector $\vec{b}$, for any α changing on the interval $0 \leq \alpha < 90^\circ$, there will exist a modified vector $\vec{c}$ conjugated to it.

In particular case, for $\alpha = 90°$, by definition of scalar product of vectors, i.e. in the case of orthogonality of the vectors $\vec{a} \perp \vec{b}$, for the vector $\vec{a}$ there will not exist a vector conjugated to it. Thus, we establish that for an arbitrary α changing on the interval $0 \leq \alpha < 90^\circ$, for the vector $\vec{a}$ adjacent to the vector $\vec{b}$, there will exist a conjugated modified vector $\vec{c}$ defined by Eq. (62).

1.3.6 EXAMPLE 2

Let in three-dimensional vector-space R_3 the following points (vectors) (Fig. 4) be given:

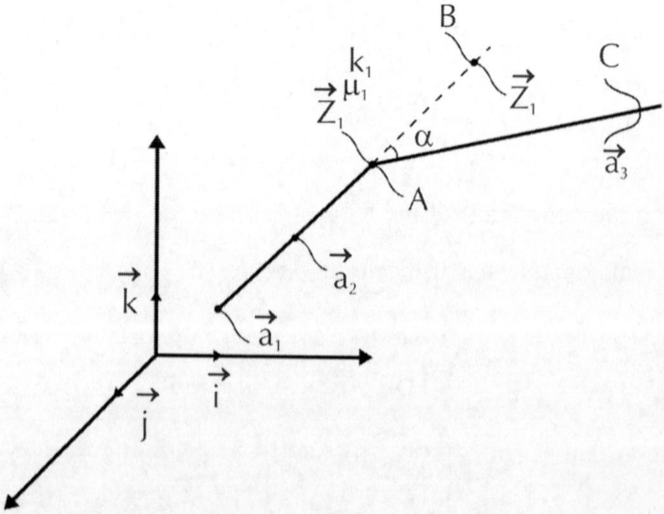

FIGURE 4 Definition of a conjugate vector.

$$\vec{a}_1 = \vec{i} + \vec{j} + \vec{k}\ ,\ \vec{a}_2 = 3\vec{i} + 2\vec{j} + 4{,}5\vec{k}\ ,$$

$$\vec{a}_3 = 6\vec{i} + 4\vec{j} + 7\vec{k} \tag{63}$$

By means of these points (vectors) write two adjacent vector equations of piecewise-linear straight lines with contiguity angle α, and then show the construction of symbolic operation of division of there vectors one by another, and also construction of a conjugated vector for one of these piecewise-linear vectors. According to the general theory of finite-dimensional Euclidean space, the vector equation of the straight line, in the general case is written in the form:

$$\vec{z}_i = \vec{a}_i + \mu_i(\vec{a}_{i+1} - \vec{a}_i) \tag{64}$$

Here the parameters μ_i accept the following values: $\vec{a} \cdot \vec{b} = |\vec{a}| \cdot |\vec{b}|$ for the points of the segment; $\mu_i \geq 0$ for the points of the straight line. By means of the points $\vec{a}_1$ and $\vec{a}_2$ in three-dimensional Euclidean space we write a vector function for the first piecewise linear straight lines in the form:

$$\vec{z}_1 = \vec{a}_1 + \mu_1(\vec{a}_2 - \vec{a}_1) \tag{65}$$

Taking into account Eq. (63), Eq. (65) in the coordinate form is written as follows:

$$\vec{z}_1 = (1+2\mu_1)\vec{i} + (1+\mu_1)\vec{j} + (1+3{,}5\mu_1)\vec{k} \tag{66}$$

Accept the parameter μ_1 for the contiguity point between piecewise-linear straight lines equal to 1.5, i.e. $\mu_1^{k_1} = 1{,}5$. For this value of the parameter $\mu_1^{k_1}$, the value of the adjoint point will equal:

$$\vec{z}_1^{\mu_1^{k_1}} = 4\vec{i} + 2{,}5\vec{j} + 6{,}25\vec{k} \tag{67}$$

Now, by means of contiguity point Eq. (67) of the point a_3 and arbitrary point $\vec{z}_1$, construct two equations of piecewise-linear straight lines $\overrightarrow{AB}$ and $\overrightarrow{AC}$ (Fig. 4).

In this case, vector equations of piecewise-linear straight lines $\overrightarrow{AB}$ and $\overrightarrow{AC}$ will take the form:

$$\overrightarrow{AB} = \vec{z}_1 - \vec{z}_1^{\mu_1^{k_1}} \tag{68}$$

$$\overrightarrow{AC} = \vec{a}_3 - \vec{z}_1^{\mu_1^{k_1}} \tag{69}$$

Allowing for Eqs. (66) and (67), Eq. (68) in coordinate form will take the form:

$$\overrightarrow{AB} = (-3+2\mu_1)\vec{i} + (-1{,}5+\mu_1)\vec{j} + (-5{,}25+3{,}5\mu_1)\vec{k} \tag{70}$$

Here the parameter $\mu_1 \geq 1{,}5$ takes the value more than 1.5, i.e., $\vec{a} = |\vec{a}| \cdot \vec{a}_0$.

For conducting appropriate numerical calculation we choose an arbitrary value for μ_1. Accept $\mu_1 = 2$. Then Eq. (70) will take the form:

$$\overrightarrow{AB} = \vec{i} + 0{,}5\vec{j} + 1{,}75\vec{k} \tag{71}$$

Substituting Eqs. (63) and (67) in Eq. (69), we write the vector $\overrightarrow{AC}$ in the coordinate form as follows:

$$\overrightarrow{AC} = 2\vec{i} + 1{,}5\vec{j} + 0{,}75\vec{k} \tag{72}$$

Now, by means of the constructed vector equations of piecewise-linear straight lines Eqs. (71) and (72) study the peculiarities of symbolic representation of the operation of division of these vectors:

$$\frac{\overrightarrow{AC}}{\overrightarrow{AB}} = \frac{2\vec{i} + 1{,}5\vec{j} + 0{,}75\vec{k}}{\vec{i} + 0{,}5\vec{j} + 1{,}75\vec{k}} \tag{73}$$

In order to get rid of the vector property in the denominator of Eq. (73), it is necessary for the vector in the numerator $\overrightarrow{AC}$ to find some modified vector $\vec{c}$ conjugated to it.

The essence of construction of the modified vector $\vec{c}$ is that the scalar product of the vector of numerator $\overrightarrow{AC}$ the modified vector $\vec{c}$ conjugated to it is a unit, i.e., $\vec{a} \cdot \vec{c} = 1$. As in Eq. (73) the piecewise-linear vectors $\overrightarrow{AB}$ and $\overrightarrow{AC}$ have a common contiguity point $\vec{z}_1^{\,\mu_1^{k_1}}$ with contiguity angle α, by means of these given two piecewise-linear vectors $\overrightarrow{AB}$ and $\overrightarrow{AC}$ by using theorem two, to construct some vector $\vec{c}$ adjacent to the numerator.

It will be defined in the form:

$$\vec{c} = \eta \cdot \overrightarrow{AB},$$

where

$$\eta = \frac{1}{\left|\overrightarrow{AB}\right| \cdot \left|\overrightarrow{AB}\right| \cos\alpha} \cdot = \frac{\left|\overrightarrow{AB}\right|\left|\overrightarrow{AC}\right| \cdot \cos\alpha}{\overrightarrow{AB}^2}$$

Now, multiplying the numerator and denominator of the vector of fraction $\frac{\overrightarrow{AC}}{\overrightarrow{AB}}$ by the conjugated vector $\vec{c}$, we get:

$$\frac{\overrightarrow{AC}}{\overrightarrow{AB}} = \frac{\overrightarrow{AC}\cdot\vec{c}}{\overrightarrow{AB}\cdot\vec{c}} = \frac{1}{\overrightarrow{AB}\cdot\vec{c}} = \frac{1}{\overrightarrow{AB}\cdot\mu\overrightarrow{AB}} = \frac{\left|\overrightarrow{AB}\right|\left|\overrightarrow{AC}\right|\cos\alpha}{\overrightarrow{AB}^2} \tag{74}$$

Then the expression of the conjugated vector $\vec{c}$ will be of the form:

$$\vec{c} = \eta(\vec{i} + 0{,}5\vec{j} + 1{,}75\vec{k}) = \\ = \frac{\vec{i} + 0{,}5\vec{j} + 2{,}75\vec{k}}{\left|\vec{i} + 0{,}5\vec{j} + 2{,}75\vec{k}\right|\cdot\left|2\vec{i} + 1{,}5\vec{j} + 1{,}75\vec{k}\right|\cdot\cos\alpha} \tag{75}$$

Applying Eq. (74) to Eq. (73), and taking into account the expression for the conjugated vector $\vec{c}$ Eq. (75), and also Eqs. (71) and (72), we define the value of the fraction vectors in the form:

$$\frac{\overrightarrow{AC}}{\overrightarrow{AB}} = \frac{\left|\vec{i} + 0{,}5\vec{j} + 1{,}75\vec{k}\right|\cdot\left|2\vec{i} + 1{,}5\vec{j} + 0{,}75\vec{k}\right|}{(\vec{i} + 0{,}5\vec{j} + 1{,}75\vec{k})^2}\cdot\cos\alpha = \\ = 1{,}2569\cdot\cos\alpha \tag{76}$$

Using the equation of scalar product of two vectors, calculate the value of $\cos\alpha$, formed between piecewise-linear straight lines $\overrightarrow{AB}$ and $\overrightarrow{AC}$:

$$cos\alpha = \frac{\overrightarrow{AC}\cdot\overrightarrow{AB}}{\left|\overrightarrow{AC}\right|\cdot\left|\overrightarrow{AB}\right|} \tag{77}$$

Substituting Eqs. (71) and (72) in Eq. (77), we get the value of *cos α*:

$$cos\alpha = \frac{(\vec{i} + 0{,}5\vec{j} + 1{,}75\vec{k})\cdot(2\vec{i} + 1{,}5\vec{j} + 0{,}75\vec{k})}{\left|\vec{i} + 0{,}5\vec{j} + 1{,}75\vec{k}\right|\cdot\left|2\vec{i} + 1{,}5\vec{j} + 0{,}75\vec{k}\right|} = \\ = \frac{4{,}0625}{5{,}4284} = 0{,}7484 \tag{78}$$

Substituting Eq. (78) in Eq. (76), we finally get:

$$\frac{\overrightarrow{AC}}{\overrightarrow{AB}} = \frac{2\vec{i} + 1{,}5\vec{j} + 0{,}75\vec{k}}{\vec{i} + 0{,}5\vec{j} + 1{,}75\vec{k}} = 0{,}9407 \tag{79}$$

1.3.7 THEOREM 3

In the Euclidean space, however dividing the collinear vectors $\vec{a}$ by $\vec{b}$, the scalar product of corresponding conjugated vectors $\vec{c}_1$ and $\vec{c}_2$ is inversely proportional to the product of the vectors $\vec{a}$ and $\vec{b}$ themselves.

1.3.8 PROOF

From theorem one it follows that by means of two collinear vectors $\vec{a}$ and $\vec{b}$ of Euclidean space, one can construct for these vectors the corresponding conjugated vectors of the form:

$$\vec{a} \cdot \vec{c}_1 = 1 \quad \vec{b} \cdot \vec{c}_2 = 1 \tag{80}$$

where the appropriate conjugated vectors $\vec{c}_1$ and $\vec{c}_2$ have the form:

$$\vec{c}_1 = \frac{1}{|\vec{a}| \cdot |\vec{b}|} \cdot \vec{b} = \eta \vec{b} \quad , \quad \vec{c}_2 = \frac{1}{|\vec{a}| \cdot |\vec{b}|} \cdot \vec{a} = \eta \vec{a} \tag{81}$$

Here η is said to be a contingency factor of the vectors $\vec{a}$ and $\vec{b}$.

Using Eqs. (80) and (81) we prove that,

$$\vec{c}_1 \cdot \vec{c}_2 = \frac{1}{\vec{a} \cdot \vec{b}} = \eta \tag{82}$$

For that we consider the vector fraction $\frac{\vec{a}}{\vec{b}}$. Multiply the numerator and denominator of this fraction by the conjugated vector of numerator $\vec{c}_1$, and taking into account Eq. (80), we get:

$$\frac{\vec{a}\cdot\vec{c}_1}{\vec{b}\cdot\vec{c}_1}=\frac{1}{\vec{b}\cdot\vec{c}_1} \tag{83}$$

Similarly, multiply the numerator and denominator of the vector fraction by appropriate conjugated vector $\vec{c}_2$ conjugated to the denominator, take into account Eq. (80), and we get:

$$\frac{\vec{a}\cdot\vec{c}_2}{\vec{b}\cdot\vec{c}_2}=\vec{a}\cdot\vec{c}_2 \tag{84}$$

For the operations Eqs. (83) and (84) hold, it is necessary that the obtained relation to give one and the same result, i.e.:

$$\frac{1}{\vec{b}\cdot\vec{c}_1}=\vec{a}\cdot\vec{c}_2 \tag{85}$$

Hence, it follows that

$$\vec{c}_1\cdot\vec{c}_2=\frac{1}{\vec{a}\cdot\vec{b}}=\eta \tag{86}$$

Q.E.D.

1.3.8 EXAMPLE 3

Let in three-dimensional space R_3 the two vectors be given:

$$\vec{a}=2\vec{i}+4\vec{j}+6\vec{k}\quad \vec{b}=\vec{i}+2\vec{j}+3\vec{k} \tag{87}$$

We prove o this example that the scalar product of the conjugated vectors $\vec{c}_1$ and $\vec{c}_2$ is inversely proportional to the scalar product of the vectors $\vec{a}$ and $\vec{b}$ themselves, i.e., equals the contingency factor of these vectors η. On the other hand, show how the conjugated vector $\vec{c}_1$ corresponding to the denominator vector is determined by means of one conjugated vector $\vec{c}_2$.

According to Theorem 1, the conjugated vector for a vector situated in the numerator of the fraction $\frac{\vec{a}}{\vec{b}}$ will equal:

$$\vec{c}_1 = \frac{1}{\vec{a} \cdot \vec{b}} \cdot \vec{b} = \eta \vec{b} \tag{88}$$

Taking into account Eq. (54), the contingency factor η of the vectors $\vec{a}$ and $\vec{b}$ will equal:

$$\eta = \frac{1}{(2\vec{i} + 4\vec{j} + 6\vec{k})(\vec{i} + 2\vec{j} + 3\vec{k})} = \frac{1}{28} \tag{89}$$

Then allowing for Eqs. (87) and (89), Eq. (88) will take the form:

$$\eta = \frac{1}{(2\vec{i} + 4\vec{j} + 6\vec{k})(\vec{i} + 2\vec{j} + 3\vec{k})} = \frac{1}{28} \tag{90}$$

According to Eq. (86) of Theorem 3, the conjugated vector $\vec{c}_2$ corresponding to the denominator vector will have the form:

$$\vec{c}_2 = \frac{\eta}{\vec{c}_1} \tag{91}$$

Substituting Eqs. (89) and (90) in Eq. (91), define the numerical value of the conjugated vector $\vec{c}_2$ in the form:

$$\vec{c}_2 = \frac{1}{28} \cdot \frac{1}{\frac{1}{28}(\vec{i} + 2\vec{j} + 3\vec{k})} = \frac{1}{\vec{i} + 2\vec{j} + 3\vec{k}} \tag{92}$$

Thus, having the value of the conjugated vector $\vec{c}_1$ for the numerator of the fraction, we defined the appropriate value of the conjugated vector $\vec{c}_2$ that will correspond to the vector in the denominator of the fraction.

The fact that the found vector $\vec{c}_2$ is a vector conjugated to the vector of the denominator of the fraction $\vec{b}$ follows from equality to a unit of the scalar product of the vectors $\vec{b}$ on $\vec{c}_2$, i.e.,

$$\vec{b}\vec{c}_2 = 1 \tag{93}$$

Taking into account Eqs. (87) and (92), we get:

$$\vec{b}\vec{c}_2 = (\vec{i} + 2\vec{j} + 3\vec{k}) \cdot \frac{1}{\vec{i} + 2\vec{j} + 3\vec{k}} \equiv 1 \qquad (94)$$

Q.E.D.

1.3.9 THEOREM 4

In Euclidean space R_n when dividing the contiguous vectors $\vec{a}$ by $\vec{b}$, the scalar product of their conjugated vectors $\vec{c}_1$ and $\vec{c}_2$ is inversely proportional to the scalar product of the vectors $\vec{a}$ and $\vec{b}$.

1.3.10 PROOF

According to Theorem 2, when dividing the adjoint vectors $\vec{a}$ and $\vec{b}$ one by another, for any contiguity angle $0 \le \alpha \le 90^0$ there will exist corresponding modified conjugated vectors $\vec{c}_1$ and $\vec{c}_2$ of the form:

$$\vec{c}_1 = \frac{1}{|\vec{a}| \cdot |\vec{b}| \cos\alpha} \cdot \vec{b} = \eta \cdot \vec{b} \qquad (95)$$

$$\vec{c}_2 = \frac{1}{|\vec{a}| \cdot |\vec{b}| \cos\alpha} \cdot \vec{a} = \eta\vec{a} \qquad (96)$$

where $\eta = \frac{1}{|\vec{a}| \cdot |\vec{b}| \cos\alpha}$ is a contingency factor between the adjoint vectors $\vec{a}$ and $\vec{b}$. And these modified conjugated vectors $\vec{c}_1$ and $\vec{c}_2$ will satisfy the conditions:

$$\vec{a} \cdot \vec{c}_1 = 1, \ \vec{b} \cdot \vec{c}_2 = 1 \qquad (97)$$

In operator of division of adjacent vectors $\frac{\vec{a}}{\vec{b}}$ we use condition Eq. (97) and prove that:

$$\vec{c}_1 \cdot \vec{c}_2 = \frac{1}{|\vec{a}| \cdot |\vec{b}| \cos\alpha} = \eta \qquad (98)$$

For that we consider the vector fraction $\frac{\vec{a}}{\vec{b}}$. Multiply the numerator and denominator of this fraction by the vector $\vec{c}_1$, conjugated to the numerator that is conjugated to the numerator vector $\vec{a}$ take into account the conjugation condition Eq. (97) and get:

$$\frac{\vec{a}}{\vec{b}} = \frac{\vec{a} \cdot \vec{c}_1}{\vec{b} \cdot \vec{c}_1} = \frac{1}{\vec{b} \cdot \vec{c}_1} \qquad (99)$$

Similarly, multiply and divide the vector fraction by the conjugated vector $\vec{c}_2$, which is conjugated to the denominator vector $\vec{b}$, take into account conjugation condition Eq. (97) and get:

$$\frac{\vec{a} \cdot \vec{c}_2}{\vec{b} \cdot \vec{c}_2} = \vec{a} \cdot \vec{c}_2 \qquad (100)$$

For the operations Eqs. (99) and (100) to hold, it is necessary that both of these expressions to give one and the same result, i.e.,

$$\frac{1}{\vec{b} \cdot \vec{c}_1} = \vec{a} \cdot \vec{c}_2 \qquad (101)$$

Hence it follows that:

$$\vec{c}_1 \cdot \vec{c}_2 = \frac{1}{|\vec{a}| \cdot |\vec{b}| \cos\alpha} = \eta \qquad (102)$$

From this condition it follows that for the given modified conjugated vector $\vec{c}_1$ from Eq. (102) we can define the modified conjugated vector $\vec{c}_2$ in the form:

$$\vec{c}_2 = \frac{\eta}{\vec{c}_1}, \ \eta = \frac{1}{|\vec{a}| \cdot |\vec{b}| \cos\alpha} \qquad (103)$$

KEYWORDS

- **adjoint vectors**
- **Bunyakovsky inequality**
- **conjugate vector**
- **Minkovsky inequality**
- **n-dimensional real Euclidean space**

REFERENCES

1. Bugrov, Ya. S.; Nikolskiy, S. M. *Elements of linear algebra and analytical geometry.* Ed. M. "Nauka," Moscow, **1980,** 175.
2. Kantarovich, L. V.; Gorstko, A. B. *Optimal decisions in economics.* Ed. M.: "Nauka," Moscow. **1979**.
3. Kantorovich, A. V.; Krylov, I. *Approximate methods of higher analysis.* Ed. M. "Fizmatlit," Moscow, **1962,** 590.
4. Salimovv, Y. S.; Rasulov, T. M. *Linear algebse and linear programming.* Ed. "Elm," Baku, **2006,** 300.
5. Solodovnikov, A. S.; Babaytsev, A.; Brailov, A. *Mathematics in economics.* Ed. M. "Finansi i statistika," Moscow, **2000,** 220.
6. Halmosh, R. *Finite-dimensional vector spaces.* Moscow, Ed. M. "Fizmatgiz," Moscow, **1963**.
7. Aliyev, R. A.; Aliyev, R. R. *Soft computing.* Ed. "Cashioglu." Baku, **2004,** 620.
8. Ashbi, R. U. *Introduction to cybernetics.* Ed. M. "Nauka," Moscow, **1959**.
9. Aliyev Azad, G. *On construction of conjugate vector in Euclidean space for economic-mathematical modeling.* Izvestia NASA Ser. of Humanitarian and social sciences (economics), Baku, **2007,** *2,* 242–246.
10. Aliyev Azad, G. *On a mathematical aspect of constructing a conjugate vector for defining the unaccounted factors influence function of economic process in finite-dimensional vector space,* "Urgent social-economic problems of oil, chemical and machine-building fields of economics" (collection of scientific papers of the faculty "International Economic relations and Management" of Azerbaijan. State Oil Academy) Baku, **2007,** 600–616.
11. Aliyev Azad, G. *Some problems of prediction and control of economic event in uncertainty conditions in 3-dimensional vector space.* Proceedings of the II International scientific-practical Conference "Youth and science: reality and future," Natural and Applied Sciences. Nevinnomysk, **2009,** *8,* 394–396.

CHAPTER 2

BASES OF PIECEWISE-LINEAR ECONOMIC-MATHEMATICAL MODELS WITH REGARD TO INFLUENCE OF UNACCOUNTED FACTORS IN FINITE-DIMENSIONAL VECTOR SPACE

CONTENTS

2.1 PRINCIPLE OF CERTAINTY OF ECONOMIC PROCESS IN FINITE-DIMENSIONAL VECTOR SPACE

Representation of economic problems in finite-dimensional vector space, in particular, in Euclidean space, at uncertainty conditions in the form of mathematical models is connected with difficulty of complete consideration of such important problems as: spatial inhomogeneity of the occurring economic processes; incomplete macro, micro and social-political information; time changeability of multifactor economic indices, their duration and velocity of their change.

The above-listed ones in mathematical plan reduce the solution of this problem to creation of very complex economic-mathematical models of nonlinear form.

In this connection, all possible economic processes considered with regard to uncertainty factor in a finite-dimensional vector space should be precisely determined in spatial-time aspect. Owing only to the formulated principle of spatial-time certainty of the economic process at uncertainty conditions in finite-dimensional vector space, the dynamics and structure of the occurring process may be revealed systematically. Besides, supposing a number of softened additional conditions on the occurring process, it is possible to classify it in finite-dimensional vector space and to suggest a new science based method of multivariant prediction of economic process and its control in finite-dimensional vector space at uncertainty conditions, in particular, with regard to influence of disregarded factors. In connection with above-stated, below We formulate the postulate spatial-time certainty of economic process at uncertainty conditions in finite-dimensional vector space."

2.1.1 PRINCIPLE OF SPATIAL-TIME CERTAINTY OF ECONOMIC PROCESS AT UNCERTAINTY CONDITIONS IN FINITE-DIMENSIONAL VECTOR SPACE

We assume that while investigating economic problems in spatial-time system, i.e., in finite-dimensional vector space, the occurring economic process possesses spatial inhomogeneity including a series of unaccounted factors of spatial form (especially, insufficient macro, micro and social-political information and etc.). This means that in different statistical points of finite-dimensional vector space, the nature of vector functions of the economic process will be different.

On the other hand, this process will be unstationary in time and this reflects changeability in time of multifactor economic indices and velocity of their change. However, at the point and small volume around the point $\Delta V_n(x_1, x_2, ... x_m)$ of finite-dimensional vector space we accept the economic process as homogeneous. This assumption allows us in the small volume $\Delta V(x_1, x_2, ... x_m)$ of finite-dimensional vector space to represent the economic process in the vector form as a piecewise-linear function. While passing from the points of one spatial vector piecewise-linear straight line to another spatial vector piecewise straight line, the occurring processes will be different by their homogeneity. Such a distinction will be a result of influence of above-mentioned unaccounted external factors that we call "influence functions of unaccounted parameters."

We will call such a basis the principle of "spatial-time certainty of the economic process at uncertainty conditions in finite-dimensional vector space" [1–11].

Thus, we assume that in finite-dimensional vector space, any economic process under consideration will be homogeneous if in the chosen small volume of the space $\Delta V_n(x_1, x_2, ... x_m)$ for a small time interval Δt the table of statistical data or experimental dependence of the vector function on spatial coordinates and time are obtained at the same external conditions, i.e., in the form:

$$\vec{z}_n = \lambda_n \vec{a}_n + \mu_n \vec{a}_{n+1}$$

This circumstance allows establishing correspondence both between piecewise-homogeneous small volumes $\Delta V_n(x_1, x_2, ... x_m)$ and $\Delta V_{n+1}(x_1, x_2, ... x_m)$ of neighboring statistical points and changes of economic process occurring in the finite-dimensional vector space.

Mathematically, this means that in the case of homogeneous process, the piecewise-linear vector-function $\vec{z}_n$ at the point and at its small vicinity $\Delta V_n(x_1, x_2, ... x_m)$ is an analytic function. The derivatives in coordinates and also velocity of change of the vector function in small time interval will be constant.

Based on this principle, we suggest a way for constructing piecewise-linear economic-mathematical models in finite-dimensional vector space with regard to influence of unaccounted factors.

2.2 GENERAL METHOD FOR CONSTRUCTING PIECEWISE-LINEAR ECONOMIC-MATHEMATICAL MODELS WITH REGARD TO INFLUENCE OF UNACCOUNTED FACTORS IN FINITE-DIMENSIONAL VECTOR SPACE

Let the statistical table describing some economic process in the form of the points (vectors) set $\{\vec{a}_n\}$ of finite-dimensional vector space R_m be given the numbers $a_{n1}, a_{n2}, a_{n3}, \ldots a_{nm}$ be the coordinates of the point (of the vector) $\vec{a}_n$ or in the name the components of the vector. By means of the vectors $\vec{a}_n$ represent all the set of statistical points (vectors) in the space R_m in the vector form in the form of n-piecewise-linear equations of the form:

$$\vec{z}_n = \lambda_n \vec{a}_n + \mu_n \vec{a}_{n+1}, \quad (\lambda_n, \mu_n \geq 0, \lambda_n + \mu_n = 1) \tag{1}$$

Here λ_n and μ_n are arbitrary positive numbers referred to the n-piece-wise-linear straight line, whose sums equals 1.

It the numbers λ_n and μ_n are nonnegative, the set of the vectors $\vec{z}_n$ of the Eq. (1) defines the equation of the n-th section in the vector form. In the case λ_n and μ_n of any sign the set of vectors $\vec{z}_n$ of the Eq. (1) will determine the equation of the n-th straight line in the vector form.

Write the vector equations of piecewise-linear straight lines Eq. (1) with regard to connections of parameters $\lambda_n + \mu_n = 1$ depending only on one parameter λ_n or μ_n in the form (Fig. 1.)

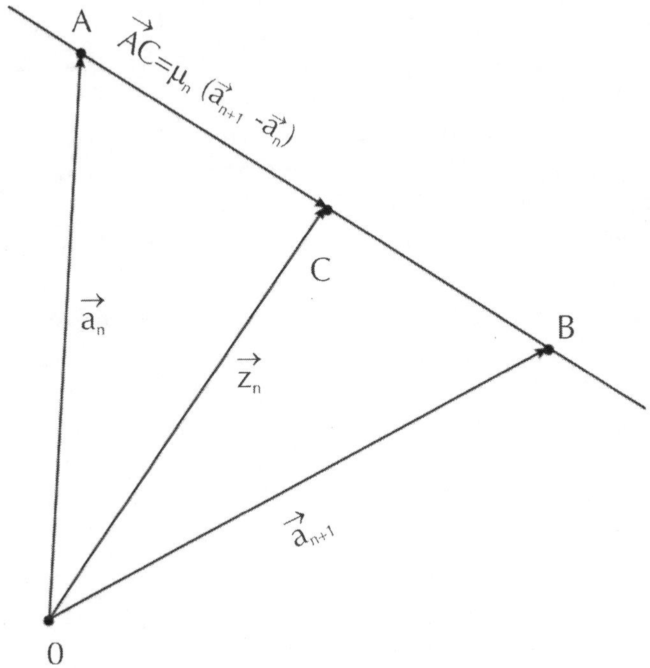

FIGURE 1 Dependence of the points (vectors) $\vec{z}_n$ of the straight line AB on two points $\vec{a}_n$ and $\vec{a}_{n+1}$ in finite-dimensional vector space R_m.

$$\vec{z}_n = \vec{a}_n + \mu_n(\vec{a}_{n+1} - \vec{a}_n),\ 0 \le \mu_n \le 1 \tag{2}$$

or

$$\vec{z}_n = \vec{a}_{n+1} + \lambda_n(\vec{a}_n - \vec{a}_{n+1}),\ 0 \le \lambda_n \le 1 \tag{3}$$

It becomes clear from equality Eq. (2) that in three-dimensional space, the points $\vec{z}_n$ fill the segment connecting the points $\vec{a}_n$ and $\vec{a}_{n+1}$. This is seen from the fact that the radius-vector $\vec{z}_n$ is the sum of the vector $\vec{a}_n$ and the vector $\mu_n(\vec{a}_{n+1} - \vec{a}_n)$, collinear with the vector $(\vec{a}_{n+1} - \vec{a}_n)$. Thus, the points set Eq. (1) is the segment $[\vec{a}_n, \vec{a}_{n+1}]$ in the space R_3, connecting the points $\vec{a}_n$ and $\vec{a}_{n+1}$. For $\mu_n = 0$, $\vec{z}_n = \vec{a}_n$; for $\lambda = 0$, $\vec{z}_n = \vec{a}_{n+1}$; for any $\lambda_n > 0$ $(\mu_n = 1 - \lambda_n > 0)$ the point $\vec{z}_n$ is an arbitrary point of the segment $[\vec{a}_n, \vec{a}_{n+1}]$.

Note that in 3-dimensional space R_3 the set point:

$$\vec{z}_n = \lambda_n \vec{a}_n + \mu_n \vec{a}_{n+1}, \quad \lambda_n + \mu_n = 1 \tag{4}$$

where λ_n and μ_n of any sign are piecewise-linear straight lines in the vector form, passing through the points $\vec{a}_n$ and $\vec{a}_{n+1}$.

In the space $R_m (m > 3)$ the points set Eq. (4) will be said a piecewise-linear straight line by definition.

Note that from the principle of spatial-time certainty of economic process at uncertainty conditions in finite-dimensional vector space it follows that in the small volume $\Delta V_n(x_1, x_2, ... x_m)$ around the point of finite-dimensional vector space each of the considered piecewise-linear vector functions $\vec{z}_n = \vec{z}_n[\omega_n(\lambda_n, \alpha_{n-1,n})]$ will be homogeneous by itself, i.e., the vector function $\vec{z}_n = \vec{z}_n[\omega_n(\lambda_n, \alpha_{n-1,n})]$ will be an analytic function, and the derivatives of this vector function in coordinates and time will be constant. And homogeneity degree of piecewise-linear vectors will be different [12–15].

This means that each piecewise-linear vector-function $\vec{z}_n$ is obtained by observing certain external factors that are inherent to the points of definite small volume $\Delta V_n(x_1, x_2, ... x_m)$ at time interval Δt_n.

Thus, economic events occurring in preceding small volumes ΔV_1, ΔV_2, ... ΔV_{n-1} will influence on the economic process in the small volume ΔV_n.

Now, using the principle of spatial-time certainty of economic process in finite-dimensional vector space we give a method for constructing the n-th piecewise-linear vector-function $\vec{z}_n = \vec{z}_n[\omega_n(\lambda_n, \alpha_{n-1,n})]$ depending on the first piecewise-linear vector equation and the cosines of the angles between the contiguous piecewise-linear vectors [14, 16–22].

We begin our construction with the second piecewise-linear vector equation. For that we behave as follows. According to Eqs. (1), (2) and (3) we write a vector equation for the first piecewise linear straight line in the form:

$$\vec{z}_1 = \lambda_1 \vec{a}_1 + \mu_1 \vec{a}_2, \text{ for } \lambda_1 + \mu_1 = 1 \tag{5}$$

or

$$\vec{z}_1 = \vec{a}_1 + \mu_1(\vec{a}_2 - \vec{a}_1) \tag{6}$$

or

$$\vec{z}_1 = \vec{a}_2 + \lambda_1(\vec{a}_1 - \vec{a}_2) \tag{7}$$

Here λ_1 and μ_1 are arbitrary numbers corresponding to the points of the first straight line (Fig. 2). Furthermore, for $\mu_1 = 0$, $\vec{z}_1 = \vec{a}_1$; for $\mu_1 = 1$, $\vec{z}_1 = \vec{a}_2$.

By the fact that the equations of the first and second straight lines do not necessarily intersect at t he point $\vec{a}_2$, i.e., the intersection point of these straight lines are also not coincide with the point $\vec{a}_2$, therefore, denote the intersection point of the first and second piecewise-linear straight lines by $\vec{z}_1^{k_1} = \vec{z}_2^{k_2}$ (Fig. 2.).

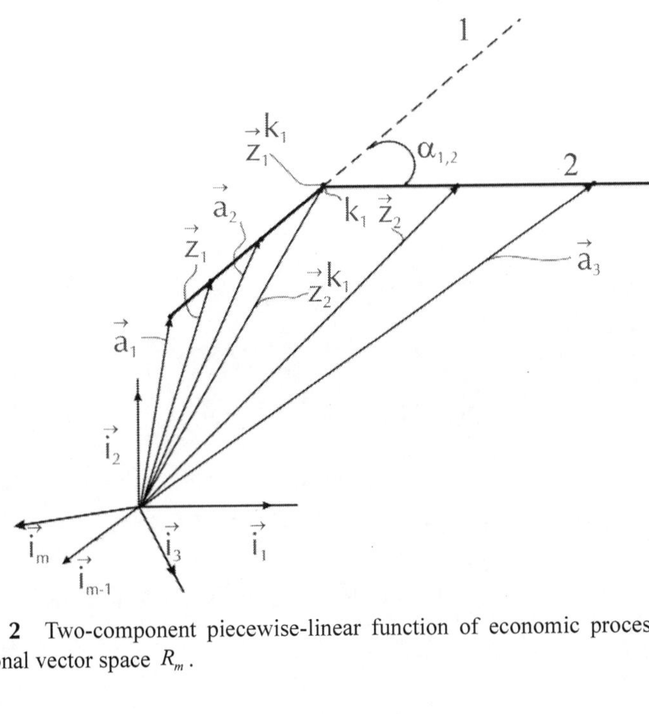

FIGURE 2 Two-component piecewise-linear function of economic process in finite-dimensional vector space R_m.

Now, by means of the intersection point $\vec{z}_1^{k_1} = \vec{z}_2^{k_1}$ and arbitrarily given point $\vec{a}_3$, write a vector equation for the second straight line in the form:

$$\vec{z}_2 = \vec{z}_2^{k_1} + \mu_2(\vec{a}_3 - \vec{z}_2^{k_1}) \tag{8}$$

Here μ_2 is an arbitrary parameter corresponding to the points of the second straight line. And for $\mu_2 = 0$, according to Eq. (8), we get the value of the intersection point of the first and second piecewise-linear straight lines, i.e., for $\mu_2 = 0$ We have $\vec{z}_2^{k_1} = \vec{z}_1^{k_1}$.

In this case, a necessary condition for the existence of the contiguity point between the straight lines will be:

$$\vec{z}_1^{k_1} = \vec{z}_2^{k_1} \tag{9}$$

Here the upper index k_1 indicates the conjugation point between the first and second piecewise-linear straight lines.

The vector equation of the second piecewise-linear straight line Eq. (8) with regard to contact condition Eq. (9) will take the form:

$$\vec{z}_2 = \vec{z}_1^{k_1} + \mu_2(\vec{a}_3 - \vec{z}_1^{k_1}) \tag{10}$$

Here $\vec{z}_1^{k_1}$ is the value of the vector (point) of the first piecewise-linear straight line at the k_1-th point. According to Eq. (6), the value $\vec{z}_1^{k_1}$ will be of the form:

$$\vec{z}_1^{k_1} = \vec{a}_1 + \mu_1^{k_1}(\vec{a}_2 - \vec{a}_1) \tag{11}$$

Here the parameter $\mu_1^{k_1}$ is the value of the parameter μ_1 in the k_1-th contact point taken from the side of the first piecewise-linear straight line.

Substituting Eqs. (11) in (10), express the vector equation of the second piecewise-linear function depending on the vectors $\vec{a}_1, \vec{a}_2, \vec{a}_3$ and the contact value of the parameter $\mu_1^{k_1}$ in the form:

$$\vec{z}_2 = [\vec{a}_1 + \mu_1^{k_1}(\vec{a}_2 - \vec{a}_1)](1 - \mu_2) + \mu_2\vec{a}_3 \tag{12}$$

Taking into account Eq. (11), we give the vector equation of the second piecewise-linear straight line Eq. (12) the following compact form:

$$\vec{z}_2 = (1 - \mu_2)\vec{z}_1^{k_1} + \mu_2 \vec{a}_3 \tag{13}$$

Thus, by Eq. (13), the points of the second piecewise-linear straight line are expressed in the vector form by the values of the conjugation point $\vec{z}_1^{k_1}$, of the vector $\vec{a}_3$ given on the second straight line and also on an arbitrary parameter μ_2 corresponding to the points of the second piecewise-linear straight line.

In particular case, if the conjugation point $\vec{z}_1^{k_1}$ coincides with the point $\vec{a}_2$, i.e., $\vec{z}_1^{k_1} = \vec{a}_2$ then in this case the parameter $\mu_1^{k_1} = 1$. By this, the vector equation of the second piecewise-linear straight line will take the following simple form:

$$\vec{z}_2 = \vec{a}_2 + \mu_2 (\vec{a}_3 - \vec{a}_2) \tag{14}$$

Now in the brackets of Eq. (12) add and subtract the quantity $\pm \mu_1 (\vec{a}_2 - \vec{a}_1)$ and also take into account the expression for $\vec{z}_1$, given by the Eq. (6). As a result we get:

$$\vec{z}_2 = \vec{z}_1 \left[1 - (\mu_1 - \mu_1^{k_1}) \frac{\vec{a}_2 - \vec{a}_1}{\vec{z}_1} + \mu_2 \frac{\vec{a}_3 - \vec{z}_1^{k_1}}{\vec{z}_1} \right]$$

for

$$\mu_1^{k_1} \le \mu_1 \le \mu_1^{k_2}, \quad 0 \le \mu_2 \le \mu_2^{k_2} \tag{15}$$

In the obtained Eq. (15), the second and third numbers are represented in the form of symtotic operation of division of vectors, more exactly, $\frac{\vec{a}_2 - \vec{a}_1}{\vec{z}_1}$ and $\frac{\vec{a}_3 - \vec{z}_1^{k_1}}{\vec{z}_1}$. However, it is known that in the axiomatic of finite-dimensional vector space, the symbolic operation of division of vectors has not been determined. Taking into account this fact, we consider this question in detail.

Here the main mathematical problem is to find the way allowing to realize the operation of division of a vector by a vector. For resolving this question, in Section 1.3 of Chapter 1 we suggest a method for constructing a conjugated vector in finite-dimensional Euclidean space. By means of the constructed conjugated vector we suggested a necessary condition that a scalar product of any vector of Euclidean space and the developed modified vector conjugated to it will equal 1. The symbolic operation of division of a vector by a vector in Euclidean space is realized by means of this condition. Below given is the applicability of this method in development of piecewise-linear economic-mathematical models in finite dimensional vector space.

Using what has been said, we have to show what will be equal the first vector fraction of Eq. (15). For that, in the vector equation of the first piecewise-linear straight line take an arbitrary piecewise vector, for instance, $(\vec{z}_1^{k_1} - \vec{a}_1)$ that will be collinear to the numerator of the vector fraction $(\vec{a}_2 - \vec{a}_1)$. Now, by means of these two vectors, i.e., of the vector $(\vec{a}_2 - \vec{a}_1)$ and of the vector $(\vec{z}_1^{k_1} - \vec{a}_1)$, and also by means of Theorem 1 of Section 1 we write the form of the conjugated modified vector. It will be in the form:

$$\vec{c} = \eta(\vec{z}_1^{k_1} - \vec{a}_1) = \frac{\vec{z}_1^{k_1} - \vec{a}_1}{|\vec{a}_2 - \vec{a}_1||\vec{z}_1^{k_1} - \vec{a}_1|} \tag{16}$$

Here

$$\eta = \frac{1}{|\vec{a}_2 - \vec{a}_1||\vec{z}_1^{k_1} - \vec{a}_1|} \tag{17}$$

is an conjugation factor between the vectors $(\vec{a}_2 - \vec{a}_1)$ and $(\vec{z}_1^{k_1} - \vec{a}_1)$.

Now, multiply the numerator and denominator of the first vector fraction of Eq. (15) by modified vector Eq. (16) conjugated to the numerator, and also take into account that the scalar product of the vector $(\vec{a}_2 - \vec{a}_1)$ and its conjugated modified vector $\vec{c} = \eta(\vec{z}_1^{k_1} - \vec{a}_1)$ equals 1, i.e.,

$$(\vec{a}_2 - \vec{a}_1)\cdot\vec{c} = (\vec{a}_2 - \vec{a}_1)\cdot\eta\cdot(\vec{z}_1^{k_1} - \vec{a}_1) = 1 \tag{18}$$

Then the first vector fraction of Eq. (15) will take the form:

$$\frac{\vec{a}_2 - \vec{a}_1}{\vec{z}_1} = \frac{(\vec{a}_2 - \vec{a}_1) \cdot \vec{c}}{\vec{z}_1 \cdot \vec{c}} = \frac{|\vec{a}_2 - \vec{a}_1||\vec{z}_1^{k_1} - \vec{a}_1|}{\vec{z}_1 (\vec{z}_1^{k_1} - \vec{a}_1)} \tag{19}$$

This will be the vector notation of the first vector fraction that will equal some scalar. The coordinate representation of (19) will look like:

$$\frac{\vec{a}_2 - \vec{a}_1}{\vec{z}_1} = \frac{\mu_1^{k_1} \sum_{m=1}^{M} (a_{2m} - a_{1m})}{\sum_{m=1}^{M} \vec{z}_{1m} \left(z_{1m}^{k_1} - a_{1m} \right)} \tag{20}$$

Similarly we show the way for getting rid of vector notation of the second vector fraction of Eq. (15). For defining an appropriate conjugated modified vector for the vector $(\vec{a}_3 - \vec{z}_1^{k_1})$, at first on the first piecewise linear straight line we choose an additional vector $(\vec{z}_1 - \vec{z}_1^{k_1})$ that will be adjacent to the vector $(\vec{a}_3 - \vec{z}_1^{k_1})$. The contiguity point of these vectors will be $\vec{z}_1^{k_1}$, and contiguity angle between these vectors will be $\alpha_{1,2}$.

Now by means of these adjacent vectors, i.e., of the vector $(\vec{a}_3 - \vec{z}_1^{k_1})$ and $(\vec{z}_1 - \vec{z}_1^{k_1})$ also of Theorem 2 of Section 1.3 we write the form of the modified vector $\vec{c}$ conjugated to the numerator. It will be of the form:

$$\vec{c} = \eta \cdot (\vec{z}_1 - \vec{z}_1^{k_1}) = \frac{\vec{z}_1 - \vec{z}_1^{k_1}}{|\vec{z}_1 - \vec{z}_1^{k_1}||\vec{a}_3 - \vec{z}_1^{k_1}| cos\alpha_{1,2}} \tag{21}$$

Here,

$$\eta = \frac{1}{|\vec{z}_1 - \vec{z}_1^{k_1}||\vec{a}_3 - \vec{z}_1^{k_1}| cos\alpha_{1,2}}$$

is the conjugation factor between adjacent vectors $(\vec{z}_1 - \vec{z}_1^{k_1})$ and $(\vec{a}_3 - \vec{z}_1^{k_1})$.

Now multiply the numerator and denominator of the second vector fraction of Eq. (15) by modified vector $\vec{c}$ of Eq. (21) conjugated to the numerator, and also take into account that the scalar product of the vector $(\vec{a}_3 - \vec{z}_1^{k_1})$ and its conjugated modified vector $\vec{c} = \eta \cdot (\vec{z}_1 - \vec{z}_1^{k_1})$ equals 1, i.e.,

$$(\vec{a}_3 - \vec{z}_1^{k_1}) \cdot \vec{c} = (\vec{a}_3 - \vec{z}_1^{k_1}) \cdot \eta \cdot (\vec{z}_1 - \vec{z}_1^{k_1}) = 1$$

Then the second vector fraction of Eq. (15) takes the form:

$$\frac{\vec{a}_3 - \vec{z}_1^{k_1}}{\vec{z}_1} = \frac{(\vec{a}_3 - \vec{z}_1^{k_1}) \cdot \vec{c}}{\vec{z}_1 \cdot \vec{c}} = \frac{\left|\vec{z}_1 - \vec{z}_1^{k_1}\right| \left|\vec{a}_3 - \vec{z}_1^{k_1}\right|}{\vec{z}_1 (\vec{z}_1 - \vec{z}_1^{k_1})} cos\alpha_{1,2} \tag{22}$$

Here, the value of $\cos\alpha_{1,2}$ between the first and second piecewise-linear straight lines is determined by means of the equation of scalar product of two adjacent vectors $\overrightarrow{AB} = \vec{z}_2 - \vec{z}_1^{k_1}$ and $\overrightarrow{AC} = \vec{z}_1 - \vec{z}_1^{k_1}$ in the form (Fig. 3):

$$\cos\alpha_{1,2} = \frac{\overrightarrow{AB} \cdot \left|\overrightarrow{AC}\right|}{\left|\overrightarrow{AB}\right| \cdot \left|\overrightarrow{AC}\right|} = \frac{(\vec{z}_1 - \vec{z}_1^{k_1})(\vec{z}_2 - \vec{z}_1^{k_1})}{\left|\vec{z}_1 - \vec{z}_1^{k_1}\right| \cdot \left|\vec{z}_2 - \vec{z}_1^{k_1}\right|}$$

While calculating the values of cos $\alpha_{1,2}$ one can use any values of arbitrary parameters μ_1 and μ_2.

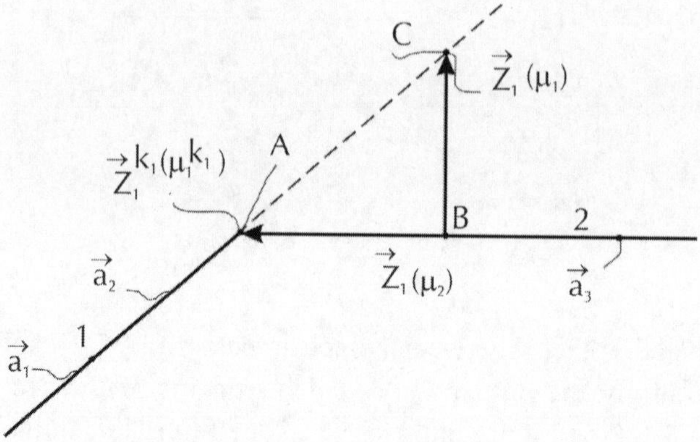

FIGURE 3 Dependence of the parameter μ_2 on the parameter μ_1 that corresponds to the points of the first piecewise-linear straight line in finite-dimensional vector space R_m.

Thus, Eq. (22) will be a vector notation of the second vector fraction of Eq. (15) that will equal some scalar. The coordinate representation Eq. (22) will have the following form:

$$\frac{\vec{a}_3 - \vec{z}_1^{k_1}}{\vec{z}_1} = \frac{\sqrt{\sum_{m=1}^{M}(a_{2m} - a_{1m})^2} \cdot \sqrt{\sum_{m=1}^{M}[a_{3m} - a_{1m} - \mu_1^{k_1}(a_{2m} - a_{1m})]^2}}{\sum_{m=1}^{M} z_{1m}(a_{2m} - a_{1m})} \cos\alpha_{1,2}$$

Substituting Eqs. (19) and (22) in Eq. (15), we get:

$$\vec{z}_2 = \vec{z}_1\left[1 + (\mu_1^{k_1} - \mu_1)\frac{|\vec{a}_2 - \vec{a}_1||\vec{z}_1^{k_1} - \vec{a}_1|}{\vec{z}_1(\vec{z}_1^{k_1} - \vec{a}_1)} + \mu_2\frac{|\vec{z}_1 - \vec{z}_1^{k_1}||\vec{a}_3 - \vec{z}_1^{k_1}|}{\vec{z}_1(\vec{z}_1 - \vec{z}_1^{k_1})}\cos\alpha_{1,2}\right] \tag{23}$$

or

$$\vec{z}_2 = \vec{z}_1\left\{1 + (\mu_1^{k_1} - \mu_1)\frac{|\vec{a}_2 - \vec{a}_1||\vec{z}_1^{k_1} - \vec{a}_1|}{\vec{z}_1(\vec{z}_1^{k_1} - \vec{a}_1)}[1 + \right.$$

$$\left. + \frac{\mu_2}{\mu_1^{k_1} - \mu_1}\frac{|\vec{z}_1 - \vec{z}_1^{k_1}||\vec{a}_3 - \vec{z}_1^{k_1}|}{\vec{z}_1(\vec{z}_1 - \vec{z}_1^{k_1})} \cdot \frac{\vec{z}_1(\vec{z}_1^{k_1} - \vec{a}_1)}{|\vec{a}_2 - \vec{a}_1||\vec{z}_1^{k_1} - \vec{a}_1|}\cos\alpha_{1,2}]\right\} \tag{24}$$

In this case, the initial point of the second piecewise linear straight line corresponding to the value of the parameters $\mu_2 = 0$ and $\mu_1 = \mu_1^{k_1} \neq 0$ will be expressed by the value of the vector-function of the first piecewise-linear straight line $\vec{z}_1$ at the point $\mu_1^{k_1}$ and by the value of the parameter $\mu_1^{k_1}$ corresponding to the intersection point, in the form:

$$\vec{z}_2\big|_{\mu_2=0} = \vec{z}_1\big|_{\mu_1=\mu_1^{k_1}}$$

In the obtained Eq. (24) the fractional expressions represent the numerical expressions dependent only on arbitrary parameters μ_1 and μ_2, changing in the interval $\mu_1^{k_1} \le \mu_1 \le \mu_1^{k_2}$ and $0 \le \mu_2 \le \mu_2^{k_2}$. Therefore, we can re-denote them in the form of numerical coefficients:

$$A = A(\mu_1) = (\mu_1^{k_1} - \mu_1)\frac{|\vec{a}_2 - \vec{a}_1||\vec{z}_1^{k_1} - \vec{a}_1|}{\vec{z}_1(\vec{z}_1^{k_1} - \vec{a}_1)}$$

$$\lambda_2 = \lambda_2(\mu_1, \mu_2) = \frac{\mu_2}{\mu_1^{k_1} - \mu_1}\frac{|\vec{z}_1 - \vec{z}_1^{k_1}||\vec{a}_3 - \vec{z}_1^{k_1}|}{\vec{z}_1(\vec{z}_1 - \vec{z}_1^{k_1})}\cdot\frac{\vec{z}_1(\vec{z}_1^{k_1} - \vec{a}_1)}{|\vec{a}_2 - \vec{a}_1||\vec{z}_1^{k_1} - \vec{a}_1|} \tag{25}$$

Here the parameter μ_1 changes in the segment $[\mu_1^{k_1}, \mu_1^{k_2}]$, the parameter μ_2 changes in the segment $[0, \mu_2^{k_2}]$, the vector $\vec{z}_1$ depends on the parameter $\mu_1 \ge \mu_1^{k_1}$ in the form:

$$\vec{z}_1 = \vec{z}_1(\mu_1).$$

Taking into account denotation Eq. (25), Eq. (24) takes the form:

$$\vec{z}_2 = \vec{z}_1\{1 + A\ (\mu_1)[1 + \lambda_2(\mu_1, \mu_2)cos\alpha_{1,2}]\}$$

for

$$\mu_1^{k_1} \le \mu_1 \le \mu_1^{k_2} \tag{26}$$

Although this equation is applicable for the case $\mu_1 \ge \mu_1^{k_2}$ as well.

Omitting the brackets in the denoted constants, we can write Eq. (26) in a more convenient form:

$$\vec{z}_2 = \vec{z}_1[1 + A\ (1 + \lambda_2 cos\alpha_{1,2})],\ \text{for}\ \mu_1^{k_1} \le \mu_1 \le \mu_1^{k_2} \tag{27}$$

Introducing a new denotation:

$$\omega_2(\lambda_2, \alpha_{1,2}) = \lambda_2 cos\alpha_{1,2} \tag{28}$$

Eq. (27) accepts finally the form:

$$\vec{z}_2 = \vec{z}_1\{1 + A[1 + \omega_2(\lambda_2, \alpha_{1,2})]\}, \text{ for } \mu_1^{k_1} \le \mu_1 \le \mu_1^{k_2} \tag{29}$$

And at the intersection point of piecewise-linear straight lines, i.e., $\mu_2 = 0$ we have:

$$\lambda_2 = 0,\ \omega_2(\lambda_2, \alpha_{1,2})\big|_{\mu_2=0} = 0$$

Thus, by Eq. (29) we mathematically establish connection of arbitrary points of the second vector equation of a piecewise-linear straight line, on one hand, depending on the equation of the first piecewise-linear straight line $\vec{z}_1$; on the other hand, from the spatial form of the parameter λ_2 and influence function $\omega_2(\lambda_2, \alpha_{1,2})$. And the parameters μ_1 and μ_2 change on the interval $\mu_1^{k_1} \le \mu_1 \le \mu_1^{k_2}$ and $0 \le \mu_2 \le \mu_2^{k_2}$.

However, it should be noted the fact that the parameters μ_1, μ_2 arbitrarily participate in Eq. (29). We can establish a special mathematical interrelation between these arbitrary parameters using, the condition of equality to zero of the scalar product of two orthogonal vectors. This allows to reduce Eq. (29) to dependence only on one arbitrary parameter. Below we have shown this procedure.

At the arbitrary point *B* of the second piecewise-linear straight line erect a perpendicular to it and continue to the intersection with the first piecewise-linear straight line. By the same token we generate the vector $\overrightarrow{BC}$ that will be perpendicular to the vector $\overrightarrow{BA}$, i.e. $\overrightarrow{BC} \perp \overrightarrow{BA}$ (Fig. 3).

In this case, we write the orthogonality condition $\overrightarrow{BC} \perp \overrightarrow{BA}$ in the form of equality to zero of their scalar product:

$$(\overrightarrow{BC} \cdot \overrightarrow{BA}) = 0 \tag{30}$$

Taking into account Eqs. (6), (10) and (11), write the expressions for the vectors $\overrightarrow{BC}$, $\overrightarrow{BA}$ and get:

$$1)\ \overrightarrow{BA} = \vec{z}_1^{k_1} - \vec{z}_2(\mu_2) = \vec{z}_1^{k_1} - [\vec{z}_1^{k_1} + \mu_2(\vec{a}_3 - \vec{z}_1^{k_1})] = -\mu_2(\vec{a}_3 - \vec{z}_1^{k_1})$$

$$\begin{aligned} 2)\ \overrightarrow{BC} &= \vec{z}_1(\mu_1) - \vec{z}_2(\mu_2) = \vec{a}_1 + \mu_1(\vec{a}_2 - \vec{a}_1) - [\vec{z}_1^{k_1} + \mu_2(\vec{a}_3 - \vec{z}_1^{k_1})] = \\ &= \vec{a}_1 + \mu_1(\vec{a}_2 - \vec{a}_1) - \{\vec{a}_1 + \mu_1^{k_1}(\vec{a}_2 - \vec{a}_1) + \mu_2[\vec{a}_3 - \vec{a}_1 - \mu_1^{k_1}(\vec{a}_2 - \vec{a}_1)]\} = \\ &= (\vec{a}_2 - \vec{a}_1)(\mu_1 - \mu_1^{k_1}) - \mu_2[\vec{a}_3 - \vec{a}_1 - \mu_1^{k_1}(\vec{a}_2 - \vec{a}_1)] \end{aligned} \tag{31}$$

Substituting Eq. (31) in scalar product Eq. (30), we get:

$$-\mu_2(\vec{a}_3 - \vec{z}_1^{k_1})\{(\vec{a}_2 - \vec{a}_1)(\mu_1 - \mu_1^{k_1}) - \mu_2[\vec{a}_3 - \vec{a}_1 - \mu_1^{k_1}(\vec{a}_2 - \vec{a}_1)]\} = 0 \tag{32}$$

From Eq. (32) it is seen that one of the values of μ_2 will be $\mu_2 = 0$. In the case $\mu_2 = 0$, the value of the parameter μ_1 will equal:

$$\mu_1 = \mu_1^{k_1} \text{ for } \mu_2 = 0 \tag{33}$$

Condition Eq. (33) will correspond to the conjugation point between the first and second piecewise-linear straight lines.

For an arbitrary value of the parameter $\mu_1 \geq \mu_1^{k_1}$ from Eq. (32) set up a contact condition between the parameters μ_2 and μ_1 in the form:

$$\mu_2 = \frac{(\vec{a}_3 - \vec{z}_1^{k_1})(\vec{a}_2 - \vec{a}_1)}{(\vec{a}_3 - \vec{z}_1^{k_1})^2}(\mu_1 - \mu_1^{k_1}), \text{ for } \mu_1^{k_1} \leq \mu_1 \leq \mu_1^{k_2} \tag{34}$$

Now, using Eq. (34), we can determine the upper value of the parameter μ_1, i.e., $\mu_1^{k_2}$. For that in Eq. (34) accept $\mu_2 = \mu_2^{k_2}$. Then from Eq. (33) we determine its appropriate value $\mu_1^{k_2}$ in the form:

$$\mu_1^{k_2} = \mu_1^{k_1} + \mu_2^{k_2}\frac{(\vec{a}_3 - z_1^{k_1})^2}{(\vec{a}_3 - z_1^{k_1})(\vec{a}_2 - \vec{a}_1)}$$

Taking into account this dependence, the range of the parameter μ_1 by defining the function of the second piecewise-linear straight line will be:

$$\mu_1^{k_1} \leq \mu_1 \leq \mu_1^{k_1} + \mu_2^{k_2} \frac{(\vec{a}_3 - z_1^{k_1})^2}{(\vec{a}_3 - z_1^{k_1})(\vec{a}_2 - \vec{a}_1)} \tag{34.a}$$

Taking into account the rule of scalar product of vectors, we write Eq. (34) in the coordinate form:

$$\mu_2 = \frac{\sum_{m=1}^{M} (a_{2m} - a_{1m})[a_{3m} - a_{1m} - \mu_1^{k_1}(a_{2m} - a_{1m})]}{\sum_{m=1}^{M} [a_{3m} - a_{1m} - \mu_1^{k_1}(a_{2m} - a_{1m})]^2} (\mu_1 - \mu_1^{k_1})$$

for $$\mu_1^{k_1} \leq \mu_1 \leq \mu_1^{k_2} \tag{35}$$

Thus, by Eqs. (34) and (35) we set up mathematical relation between the arbitrary parameters μ_2 and μ_1 corresponding to the points of the first and second piecewise-linear straight lines both in the vector and in coordinate form, respectively.

In the particular case, for $\mu_1^{k_1} = 1$, i.e., for $\vec{z}_1^{\mu_1^{k_1}} = \vec{a}_2$, Eqs. (34) and (35) will take a rather simpler form:

$$\mu_2 = \frac{(\vec{a}_2 - \vec{a}_1)(\vec{a}_3 - \vec{a}_2)}{(\vec{a}_3 - \vec{a}_2)^2} (\mu_1 - 1), \text{ for } \mu_1 \geq 1 \tag{36}$$

And in the coordinate form:

$$\mu_2 = \frac{\sum_{m=1}^{M} (a_{3m} - a_{2m})(a_{2m} - a_{1m})}{\sum_{m=1}^{M} (a_{3m} - a_{2m})^2} (\mu_1 - 1), \text{ for } \mu_1 \geq 1 \tag{37}$$

We write a vector equation for the second piecewise-linear straight line Eq. (29) in the vector form. Recall that the vectors $\vec{a}_1$, $\vec{a}_2$, $\vec{a}_3$, $\vec{z}_1$, $\vec{z}_2$, $\vec{z}_1^{k_1}$ were written in m-dimensional Euclidean space. In the coordinate form they will look like as:

$$\vec{a}_1 = \sum_{m=1}^{M} a_{1m}\vec{i}_m = a_{11}\vec{i}_1 + a_{12}\vec{i}_2 + \dots\dots\dots + a_{1M}\vec{i}_M$$

$$\vec{a}_2 = \sum_{m=1}^{M} a_{2m}\vec{i}_m = a_{21}\vec{i}_1 + a_{22}\vec{i}_2 + \dots\dots\dots + a_{2M}\vec{i}_M$$

$$\vec{a}_3 = \sum_{m=1}^{M} a_{3m}\vec{i}_m = a_{31}\vec{i}_1 + a_{32}\vec{i}_2 + \dots\dots\dots + a_{3M}\vec{i}_M$$

$$\vec{z}_1 = \sum_{m=1}^{M} z_{1m}\vec{i}_m = z_{11}\vec{i}_1 + z_{12}\vec{i}_2 + \dots\dots\dots + z_{1M}\vec{i}_M$$

$$\vec{z}_2 = \sum_{m=1}^{M} z_{2m}\vec{i}_m = z_{21}\vec{i}_1 + z_{22}\vec{i}_2 + \dots\dots\dots + z_{2M}\vec{i}_M$$

$$\vec{z}_1^{k_1} = \sum_{m=1}^{M} z_{1m}^{k_1}\vec{i}_m = z_{11}^{k_1}\vec{i}_1 + z_{12}^{k_1}\vec{i}_2 + \dots\dots\dots + z_{1mM}^{k_1}\vec{i}_M \tag{38}$$

Here $\vec{i}_m$ are unit vectors of m-dimensional space.

Taking into account Eq. (38), in the following coordinate form we write the following expressions of scalar products in of vectors and vector module:

$$1)\ |\vec{a}_2 - \vec{a}_1| = \sqrt{\sum_{m=1}^{M}(a_{2m} - a_{1m})^2}$$

$$2)\ |\vec{z}_1^{k_1} - \vec{a}_1| = \sqrt{\sum_{m=1}^{M}(z_{1m}^{k_1} - a_{1m})^2} =$$

$$= \sqrt{\sum_{m=1}^{M}\{a_{1m} + \mu_1^{k_1}(a_{2m} - a_{1m}) - a_{1m}\}^2} = \mu_1^{k_1}\sqrt{\sum_{m=1}^{M}(a_{2m} - a_{1m})^2}$$

$$3)\ |\vec{z}_1 - \vec{z}_1^{k_1}| = \sqrt{\sum_{m=1}^{M}(z_{1m} - z_{1m}^{k_1})^2} = \sqrt{(\mu_1 - \mu_1^{k_1})^2\sum_{m=1}^{M}(a_{2m} - a_{1m})^2} =$$

$$= (\mu_1 - \mu_1^{k_1})\sqrt{\sum_{m=1}^{M}(a_{2m} - a_{1m})^2}$$

$$4)\ |\vec{a}_3 - \vec{z}_1^{k_1}| = \sqrt{\sum_{m=1}^{M}(a_{3m} - z_{1m}^{k_1})^2} =$$

$$= \sqrt{\sum_{m=1}^{M}\{a_{3m} - [a_{1m} + \mu_1^{k_1}(a_{2m} - a_{1m})]\}^2}$$

$$5)\ \vec{z}_1(\vec{z}_1^{k_1} - \vec{a}_1) = \sum_{m=1}^{M} z_{1m}\mu_1^{k_1}(a_{2m} - a_{1m}) = \mu_1^{k_1}\sum_{m=1}^{M} z_{1m}(a_{2m} - a_{1m})$$

$$6)\ \vec{z}_1(\vec{z}_1 - \vec{z}_1^{k_1}) = \sum_{m=1}^{M} z_{1m}(z_{1m} - \vec{z}_1^{k_1}) =$$

$$= \sum_{m=1}^{M}(\mu_1 - \mu_1^{k_1})\sum_{m=1}^{M} z_{1m}(a_{2m} - a_{1m}) \tag{39}$$

Substituting Eq. (39) in Eq. (29) and taking into account Eq. (35), we write a vector equation for the points of the second piecewise-linear straight-line in the coordinate form:

$$\vec{z}_2 = \vec{z}_1[1 + A\,[1 + \omega_2(\lambda_2, \alpha_{12})]\}\,, \text{ for } \mu_1 \geq \mu_1^{k_1} \tag{40}$$

Here

$$A = (\mu_1^{k_1} - \mu_1)\frac{\sum_{m=1}^{M}(a_{2m} - a_{1m})^2}{\sum_{m=1}^{M} z_{1m}(a_{2m} - a_{1m})},$$

$$\lambda_2 = \frac{\mu_2}{\mu_1^{k_1} - \mu_1} \cdot \frac{\sqrt{\sum_{m=1}^{M}\{a_{3m} - [a_{1m} + \mu_1^{k}(a_{2m} - a_{1m})]\}^2}}{\sqrt{\sum_{m=1}^{M}(a_{2m} - a_{1m})^2}},$$

$$\omega_2(\lambda_2, \alpha_{1,2}) = \lambda_2 cos\alpha_{1,2},$$

$$\mu_2 = \frac{\sum_{m=1}^{M}(a_{2m} - a_{1m})[a_{3m} - a_{1m} - \mu_1^{k_1}(a_{2m} - a_{1m})]}{\sum_{m=1}^{M}[a_{3m} - a_{1m} - \mu_1^{k_1}(a_{2m} - a_{1m})]^2}(\mu_1 - \mu_1^{k_1}),$$

for

$$\mu_1 \geq \mu_1^{k_1} \tag{41}$$

We should underline the following points. The parameter λ_2 depends on an arbitrary value of the parameter μ_2 that in its turn by Eq. (36) depends on the parameter μ_1. Therefore, defining by Eq. (40) the points of the second piecewise-linear straight line, it is necessary to give the value of the parameter $\mu_1 \geq \mu_1^{k_1}$ and by means of Eq. (36) to find its appropriate parameter. After that, this value of the parameter μ_2 should be substituted in Eq. (40) and the points of the second piecewise-linear straight line depending on the parameter μ_2 corresponding to the appropriate second piecewise-linear straight line be defined.

Taking into account that in finite-dimensional vector space it holds the expression of the vectors $\vec{z}_2 = \sum_{m=1}^{M} z_{2m}\vec{i}_m$ and $\vec{z}_1 = \sum_{m=1}^{M} z_{nm}\vec{i}_m$, then according to Eq. (40) the coordinates of the vector $\vec{z}_2$, i.e., z_{2m} will be expressed by the coordinates of the first piecewise-linear vector z_{1m}, spatial parameter λ_2 and the influence function $\omega_2(\lambda_2, \alpha_{1,2})$ in the form:

$$z_{2m} = [1 + A\ (1 + \omega_2(\lambda_2, \alpha_{1,2}))]z_{1m},\ \text{for}\ m = 1,2,3,\ldots, M \tag{42}$$

Here the coefficients λ_2 and $\omega_2(\lambda_2, \alpha_{1,2})$ have the form of Eq. (41).

From Eq. (42) it is seen that by giving coordinates of the first piecewise-linear straight line z_{1m}, and also the angle α_{12} between the piecewize linear straight lines, by means of Eq. (42) the coordinates of the points of the second piecewise straight line are determined automatically.

Now, by means of the intersection point $\vec{z}_3^{\,k_2}$ and arbitrarily given point (vector) $\vec{a}_4$ taken on the third piecewise-linear vector line, we write a vector function for the third piecewise-linear straight line in the form (Fig. 4):

$$\vec{z}_3 = \vec{z}_3^{\,k_2} + \mu_3(\vec{a}_4 - \vec{z}_3^{\,k_2}) \tag{43}$$

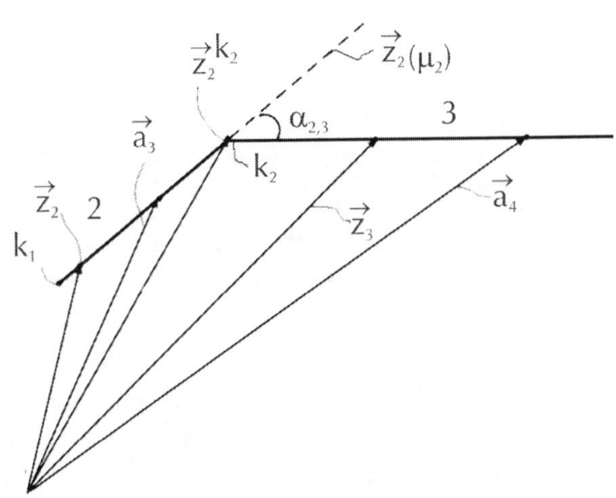

FIGURE 4 The scheme of construction of the third piecewise-linear straight line in finite-dimensional vector space R_m.

Here μ_3 is an arbitrary parameter corresponding to the points of the third straight line. And for $\mu_3 = 0$, according to Eq. (43) we get the value of the intersection point of the second and third piecewise-linear straight lines, i.e., for $\mu_3 = 0$ we have:

$$\vec{z}_3 = \vec{z}_3^{k_2}$$

In this case, the condition of existence of the adjacent point between the second and third piecewise-lines straight linear will be:

$$\vec{z}_3^{k_2} = \vec{z}_2^{k_2} \tag{44}$$

Taking into account contact condition Eq. (44), the equation for the third piecewise-linear straight line in vector form Eq. (43) will accept the form:

$$\vec{z}_3 = \vec{z}_2^{k_2} + \mu_3(\vec{a}_4 - \vec{z}_2^{k_2}) \tag{45}$$

Here $\vec{z}_2^{k_2}$ is the value of the point of the second piecewise-linear straight line at the k_2-th conjugation point.

In Eq. (45) the point (vector) $\vec{z}_2^{k_2}$ is calculated by means of the earlier derived Eq. (15). Therefore, in Eq. (15), accepting instead of μ_2 the value of $\mu_2^{k_2}$, the value of the vector $\vec{z}_2^{k_2}$ at the k_2-th point will be of the form:

$$\vec{z}_2^{k_2} = \vec{z}_1\left[1 - (\mu_1 - \mu_1^{k_1})\frac{\vec{a}_2 - \vec{a}_1}{\vec{z}_1} + \mu_2^{k_2}\frac{\vec{a}_3 - \vec{z}_1^{k_1}}{\vec{z}_1}\right] \text{ for } \mu_1^{k_1} \le \mu_1 \le \mu_1^{k_2}$$

Here the parameter $\mu_1^{k_2}$ is connected with the parameter $\mu_2^{k_2}$ in the form:

$$\mu_1^{k_2} = \mu_1^{k_1} + \mu_2^{k_2}\frac{(\vec{a}_3 - \vec{a}_2)^2}{(\vec{a}_3 - \vec{z}_1^{k_1})(\vec{a}_2 - \vec{a}_1)} \tag{46}$$

The equation of the second straight line passing through the two given conjugation points k_1 and k_2 to which there will correspond the values of the parameters $\mu_1^{k_1}$ and $\mu_2^{k_2}$ will be set up by Eq. (46). And in this case, $\vec{z}_2^{k_2}(\mu_1)$ will depend on the parameter μ_1.

Further, substitute Eq. (46) in Eq. (45) and get:

$$\vec{z}_3 = \vec{z}_1 \left[1 - (\mu_1 - \mu_1^{k_1}) \frac{\vec{a}_2 - \vec{a}_1}{\vec{z}_1} + \mu_2^{k_2} \frac{\vec{a}_3 - \vec{z}_1^{k_1}}{\vec{z}_1} \right] (1 - \mu_3) + \mu_3 \vec{a}_4 =$$
$$= \vec{z}_2^{k_2} (1 - \mu_3) + \mu_3 \vec{a}_4$$

or

$$\vec{z}_3 = \vec{z}_1 \left[1 + (\mu_1^{k_1} - \mu_1) \frac{\vec{a}_2 - \vec{a}_1}{\vec{z}_1} + \mu_2^{k_2} \frac{\vec{a}_3 - \vec{z}_1^{k_1}}{\vec{z}_1} \right] -$$

$$- \mu_3 \vec{z}_1 \frac{\vec{z}_1^{k_1} + \mu_2^{k_2} (\vec{a}_3 - \vec{z}_1^{k_1}) - \vec{a}_4}{\vec{z}_1} \quad (47)$$

Now, according to Eq. (13) the value of the point $\vec{z}_2^{k_2}$ for $\mu_2 = \mu_2^{k_2}$ will be of the form:

$$\vec{z}_2^{k_2} = (1 - \mu_2^{k_2}) \vec{z}_1^{k_1} + \mu_2^{k_2} \vec{a}_3$$

Substituting this value in (47), we get:

$$\vec{z}_3 = \vec{z}_1 \left[1 + (\mu_1^{k_1} - \mu_1) \frac{\vec{a}_2 - \vec{a}_1}{\vec{z}_1} + \mu_2^{k_2} \frac{\vec{a}_3 - \vec{z}_1^{k_1}}{\vec{z}_1} + \mu_3 \frac{\vec{a}_4 - \vec{z}_2^{k_2}}{\vec{z}_1} \right]$$

for $$\mu_1 \geq \mu_1^{k_2} \quad (48)$$

In Eq. (48) the numbers are represented in the form of symbolic operation of division of vectors. Above we showed the way for getting rid of the vector notation of the first two fraction expressions. The results of the first two fraction expressions were represented in the form of Eqs. (19) and (22).

Now get rid of the vector notation of the third fraction $\frac{\vec{a}_4 - \vec{z}_2^{k_2}}{\vec{z}_1}$. For that construct a modified vector $\vec{c}$ conjugate to the vector $(\vec{a}_4 - \vec{z}_2^{k_2})$. For that, on the vector equation of the second piecewise-linear straight line choose an additional piecewise-linear vector $(\vec{z}_2 - \vec{z}_2^{k_2})$ for $\mu_2 > \mu_2^{k_2}$ that will be adjacent to the vector $(\vec{a}_4 - \vec{z}_2^{k_2})$. The contiguity point of these vectors will be $\vec{z}_2^{k_2}$, and the adjacent angle between these vectors will be $\alpha_{2,3}$. And this angle $\alpha_{2,3}$ will lie in the plane of these vectors. Now, by means of these adjacent vectors, i.e., of the vectors $(\vec{a}_4 - \vec{z}_2^{k_2})$, and $(\vec{z}_2 - \vec{z}_2^{k_2})$, also of Theorem 2 of Section 3.1 of Chapter 1 we write the form of the conjugate modified vector $\vec{c}$. It will have the form:

$$\vec{c} = \eta \cdot (\vec{z}_2 - \vec{z}_2^{k_2}) = \frac{\vec{z}_2 - \vec{z}_2^{k_2}}{\left|\vec{z}_2 - \vec{z}_2^{k_2}\right| \left|\vec{a}_4 - \vec{z}_2^{k_2}\right| cos\alpha_{2,3}} \tag{49}$$

Here

$$\eta = \frac{1}{\left|\vec{z}_2 - \vec{z}_2^{k_2}\right| \left|\vec{a}_4 - \vec{z}_2^{k_2}\right| cos\alpha_{2,3}}$$

of is the coefficient between the adjacent vectors $(\vec{z}_2 - \vec{z}_2^{k_2})$ and $(\vec{a}_4 - \vec{z}_2^{k_2})$.

Here the value of $cos\alpha_{2,3}$ between the second and third piecewise-linear straight lines is determined by means of the scalar product of two adjacent vectors $\overrightarrow{AB} = \vec{z}_3(\mu_3) - \vec{z}_2^{k_2}(\mu_2^{k_2})$ and $\overrightarrow{AC} = \vec{z}_2(\mu_2) - \vec{z}_2^{k_2}(\mu_2)$ in the form (Fig. 5):

$$cos\alpha_{2,3} = \frac{\overrightarrow{AB} \cdot \left|\overrightarrow{AC}\right|}{\left|\overrightarrow{AB}\right| \cdot \left|\overrightarrow{AC}\right|} = \frac{[\vec{z}_3(\mu_3) - \vec{z}_2^{k_2}][\vec{z}_2(\mu_2) - \vec{z}_2^{k_2}]}{\left|\vec{z}_3(\mu_3) - \vec{z}_2^{k_2}\right| \cdot \left|\vec{z}_2(\mu_2) - \vec{z}_2^{k_2}\right|}$$

Now multiply the numerator and denominator of the third vector fraction of Eq. (48) by the modified vector $\vec{c}$ Eq. (49) conjugate to the vector $(\vec{a}_4 - \vec{z}_2^{k_2})$, and also take into account that the scalar product of the vector $(\vec{a}_4 - \vec{z}_2^{k_2})$ and the modified vector $\vec{c} = \eta \cdot (\vec{z}_2 - \vec{z}_2^{k_2})$ conjugated to it, equals 1, i.e.,

$$(\vec{a}_4 - \vec{z}_2^{k_2}) \cdot \vec{c} = (\vec{a}_4 - \vec{z}_2^{k_2}) \cdot \eta \cdot (\vec{z}_2 - \vec{z}_2^{k_2}) = 1$$

Then the third vector fraction of Eq. (48) takes the form:

$$\frac{\vec{a}_4 - \vec{z}_2^{k_2}}{\vec{z}_1} = \frac{(\vec{a}_4 - \vec{z}_2^{k_2})\cdot \vec{c}}{\vec{z}_1 \cdot \vec{c}} = \frac{\left|\vec{z}_2 - \vec{z}_2^{k_2}\right|\left|\vec{a}_4 - \vec{z}_2^{k_2}\right|}{\vec{z}_1\,(\vec{z}_2 - \vec{z}_2^{k_2})} \cos\alpha_{2,3} \tag{50}$$

This will be the vector notation of the third vector fraction of Eq. (48) that will be equal to some scalar. The coordinate notation of Eq. (50) will be of the form:

$$\frac{\vec{a}_4 - \vec{z}_2^{k_2}}{\vec{z}_1} = \frac{\sqrt{\sum_{m=1}^{M}(z_{2m} - z_{2m}^{k_2})^2} \cdot \sqrt{\sum_{m=1}^{M}(a_{4m} - z_{2m}^{k_2})^2}}{\sum_{m=1}^{M} z_{1m}(z_{2m} - z_{2m}^{k_2})} \cdot \cos\alpha_{2,3}$$

Substituting Eqs. (50), (19) and (22) in Eq. (48), write the vector equation for the points of the third piecewise-linear straight line in the form:

$$\vec{z}_3 = \vec{z}_1 \left\{ 1 + (\mu_1^{k_1} - \mu_1)\frac{\left|\vec{a}_2 - \vec{a}_1\right|\left|\vec{z}_1^{k_1} - \vec{a}_1\right|}{\vec{z}_1(\vec{z}_1^{k_1} - \vec{a}_1)} + \mu_2^{k_2}\frac{\left|\vec{z}_1 - \vec{z}_1^{k_1}\right|\left|\vec{a}_3 - \vec{z}_1^{k_1}\right|}{\vec{z}_1(\vec{z}_1 - \vec{z}_1^{k_1})}\cos\alpha_{1,2} + \right.$$

$$\left. + \mu_3 \cdot \frac{\left|\vec{z}_2 - \vec{z}_2^{k_2}\right|\left|\vec{a}_4 - \vec{z}_2^{k_2}\right|}{\vec{z}_1(\vec{z}_2 - \vec{z}_2^{k_2})}\cos\alpha_{2,3} \right\}, \text{ for } \mu_1 \geq \mu_1^{k_2},\ \mu_3 \geq 0 \tag{51}$$

or

$$\vec{z}_3 = \vec{z}_1 \left\{ 1 + (\mu_1^{k_1} - \mu_1)\frac{\left|\vec{a}_2 - \vec{a}_1\right|\left|\vec{z}_1^{k_1} - \vec{a}_1\right|}{\vec{z}_1(\vec{z}_1^{k_1} - \vec{a}_1)} \left[1 + \right.\right.$$

$$+ \frac{\mu_2^{k_2}}{\mu_1^{k_1} - \mu_1} \cdot \frac{\left|\vec{z}_1 - \vec{z}_1^{k_1}\right|\left|\vec{a}_3 - \vec{z}_1^{k_1}\right|}{\vec{z}_1(\vec{z}_1 - \vec{z}_1^{k_1})} \cdot \frac{\vec{z}_1(\vec{z}_1^{k_1} - \vec{a}_1)}{\left|\vec{a}_2 - \vec{a}_1\right|\left|\vec{z}_1^{k_1} - \vec{a}_1\right|}\cos\alpha_{1,2} +$$

$$\left.\left. + \frac{\mu_3}{\mu_1^{k_1} - \mu_1} \cdot \frac{\left|\vec{z}_2 - \vec{z}_2^{k_2}\right|\left|\vec{a}_4 - \vec{z}_2^{k_2}\right|}{\vec{z}_1(\vec{z}_2 - \vec{z}_2^{k_2})} \cdot \frac{\vec{z}_1(\vec{z}_1^{k_1} - \vec{a}_1)}{\left|\vec{a}_2 - \vec{a}_1\right|\left|\vec{z}_1^{k_1} - \vec{a}_1\right|}\cos\alpha_{2,3} \right] \right\}$$

for

$$\mu_1 \geq \mu_1^{k_2}, \mu_3 \geq 0 \tag{52}$$

Note that all the points of the third piecewise-linear straight line in *m*-dimensional vector space will be determined by Eq. (52). And the case $\mu_3 = 0$ will correspond to the value of the initial point of the third piecewise-linear straight line that will be defined by the vector function of the first piecewise-linear straight line $\vec{z}_1$, the values of the parameters $\mu_1^{k_1}$ and $\mu_2^{k_2}$ of intersection points of piecewise-linear straight lines, and also on $cos\ \alpha_{1,2}$ generated between the first and second piecewise-linear straight lines. It has the form:

$$\vec{z}_3\Big|_{\mu_3=0} = \vec{z}_1\left\{1+(\mu_1^{k_1}-\mu_1)\frac{|\vec{a}_2-\vec{a}_1||\vec{z}_1^{k_1}-\vec{a}_1|}{\vec{z}_1(\vec{z}_1^{k_1}-\vec{a}_1)}\left[1+\right.\right.$$

$$\left.\left.+\frac{\mu_2^{k_2}}{\mu_1^{k_1}-\mu_1}\cdot\frac{|\vec{z}_1-\vec{z}_1^{k_1}||\vec{a}_3-\vec{z}_1^{k_1}|}{\vec{z}_1(\vec{z}_1-\vec{z}_1^{k_1})}\cdot\frac{\vec{z}_1(\vec{z}_1^{k_1}-\vec{a}_1)}{|\vec{a}_2-\vec{a}_1||\vec{z}_1^{k_1}-\vec{a}_1|}\right]cos\alpha_{1,2}\right\}$$

Using Eq. (29), calculate the following expressions:

$$\begin{aligned}
&1)\ |\vec{z}_2-\vec{z}_2^{k_2}| = |A\vec{z}_1[\omega_2(\lambda_2,\alpha_{1,2})-\omega_2(\lambda_2^{k_2},\alpha_{1,2})]| \\
&2)\ |\vec{a}_4-\vec{z}_2^{k_2}| = |\vec{a}_4-\vec{z}_1[1+A(1+\omega_2(\lambda_2^{k_2},\alpha_{1,2}))]| \\
&3)\ (\vec{z}_2-\vec{z}_2^{k_2}) = \vec{z}_1[1+A(1+\omega_2(\lambda_2,\alpha_{1,2}))] - \\
&-\vec{z}_1[1+A(1+\omega_2(\lambda_2^{k_2},\alpha_{1,2}))] = \\
&= A\vec{z}_1[\omega_2(\lambda_2,\alpha_{1,2})-\omega_2(\lambda_2^{k_2},\alpha_{1,2})]
\end{aligned} \tag{53}$$

Here the form of the expression of the unaccounted factors parameter $\lambda_2^{k_2}$ is the expression of the form λ_2 given by Eq. (25), where the parameter μ_2 was replaced by the parameter $\mu_2^{k_2}$, i.e.:

$$\lambda_2^{k_2} = \frac{\mu_2^{k_2}}{\mu_1^{k_1} - \mu_1} \cdot \frac{\left|\vec{z}_1 - \vec{z}_1^{k_1}\right|\left|\vec{a}_3 - \vec{z}_1^{k_1}\right|}{\vec{z}_1(\vec{z}_1 - \vec{z}_1^{k_1})} \cdot \frac{\vec{z}_1(\vec{z}_1^{k_1} - \vec{a}_1)}{\left|\vec{a}_2 - \vec{a}_1\right|\left|\vec{z}_1^{k_1} - \vec{a}_1\right|}$$

$$\omega_2(\lambda_2^{k_2}, \alpha_{1,2}) = \lambda_2^{k_2} \cos\alpha_{1,2} \tag{54}$$

$$A = (\mu_1^{k_1} - \mu_1)\frac{\left|\vec{a}_2 - \vec{a}_1\right|\left|\vec{z}_1^{k_1} - \vec{a}_1\right|}{\vec{z}_1(\vec{z}_1^{k_1} - \vec{a}_1)}$$

Taking into account Eq. (54), vector function for the third piecewise-linear straight line Eq. (52) will be of the form:

$$\vec{z}_3 = \vec{z}_1\{1 + A\ [(1 + \omega_2(\lambda_2^{k_2}, \alpha_{1,2}) + \omega_3(\lambda_3, \alpha_{2,3})]\}$$

or

$$\mu_1 \geq \mu_1^{k_2} \tag{55}$$

Here

$$\omega_3(\lambda_3, \alpha_{2,3}) = \lambda_3 \cos\alpha_{2,3} =$$

$$= \frac{\mu_3}{\mu_1^{k_1} - \mu_1} \cdot \frac{\left|\vec{z}_2 - \vec{z}_2^{k_2}\right|\left|\vec{a}_4 - \vec{z}_2^{k_2}\right|}{\vec{z}_1(\vec{z}_2 - \vec{z}_2^{k_2})} \cdot \frac{\vec{z}_1(\vec{z}_1^{k_1} - \vec{a}_1)}{\left|\vec{a}_2 - \vec{a}_1\right|\left|\vec{z}_1^{k_1} - \vec{a}_1\right|} \cos\alpha_{2,3}$$

for

$$\mu_1 \geq \mu_1^{k_2} \tag{56}$$

And at the intersection point of the second and third piecewise-linear straight lines, i.e., for $\mu_3 = 0$ we have:

$$\lambda_3 = 0, \quad \omega_3(\lambda_3, \alpha_{2,3}) = 0$$

In order to write the unaccounted parameters spatial function $\omega_3(\lambda_3, \alpha_{2,3})$ by the values of the first vector $\vec{z}_1$ and $\vec{z}_1^{k_1}$ in Eq. (56), it is necessary to express the vector-functions $\vec{z}_2$ and $\vec{z}_2^{k_2}$ by means of Eqs. (25) and (29) by $\vec{z}_1$ and $\vec{z}_1^{k_1}$. Performing this operation, represent the function $\omega_3(\lambda_3, \alpha_{2,3})$ in the form:

$$\omega_3(\lambda_3, \alpha_{2,3}) = \lambda_3 \cos\alpha_{2,3} =$$

$$= \frac{\mu_3}{\mu_1^{k_1} - \mu_1} \cdot \frac{|\vec{z}_2 - \vec{z}_2^{k_2}||\vec{a}_4 - \vec{z}_2^{k_2}|}{\vec{z}_1(\vec{z}_2 - \vec{z}_2^{k_2})} \cdot \frac{\vec{z}_1(\vec{z}_1^{k_1} - \vec{a}_1)}{|\vec{a}_2 - \vec{a}_1||\vec{z}_1^{k_1} - \vec{a}_1|} \cos\alpha_{2,3}$$

$$\cdot \frac{\vec{z}_1(\vec{z}_1^{k_1} - \vec{a}_1)}{|\vec{a}_2 - \vec{a}_1||\vec{z}_1^{k_1} - \vec{a}_1|} \cos\alpha_{2,3} \quad (57)$$

Thus, by Eq. (55) we mathematically set up the relation of arbitrary points of the third vector equation of piecewise-linear straight line depending on the vector equation of the first piecewise-linear straight line, of spatial parameters $\lambda_2^{k_2}$ and λ_3 and also on the unaccounted parameter spatial influence function $\omega_3(\lambda_3, \alpha_{2,3})$ Eq. (56).

$$(\overrightarrow{BC} \cdot \overrightarrow{BA}) = 0 \quad (58)$$

Note that in Eq. (55) the parameters μ_1 and μ_3 participate arbitrarily. Below we have established mathematical relation between these parameters. At the arbitrary point ***B*** of the third piecewise-linear straight line, erect a perpendicular to it and continue it to the intersection with the second piecewise-linear straight line. Thus, we form the vector ***BC*** with its origin at the point ***B*** on the third straight line and the finite point ***C*** on the second straight line. By the same token, we geometrically form the vector $\overrightarrow{BC}$ that will be perpendicular to the vector $\overrightarrow{BA}$, i.e., $\overrightarrow{BC} \perp \overrightarrow{BA}$ (Fig. 5).

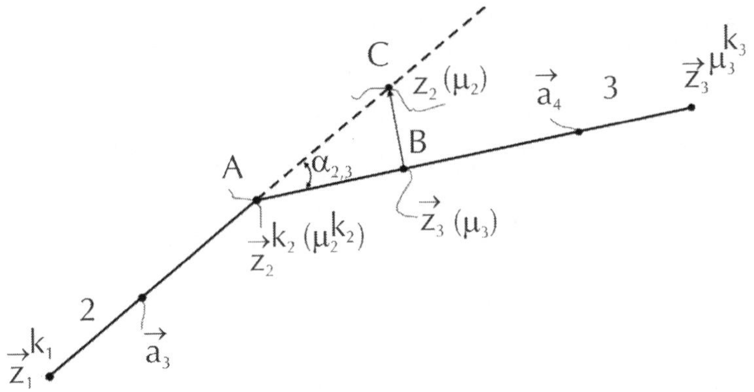

FIGURE 5 Dependence of the parameter μ_3 on the parameter μ_2 that corresponds to the points of the second piecewise-linear straight line in finite-dimension vector space R_m.

In this case, we write the orthogonality condition $\overrightarrow{BC} \perp \overrightarrow{BA}$ in the form of equality to zero of their scalar product (Fig. 5).

According to Fig. 5, we have

1. $$\vec{z}_3 = z_2^{k_2} + \mu_3(\vec{a}_4 - \vec{z}_2^{k_2}) \quad \text{for} \quad z_3^{k_2} = \vec{z}_2^{k_2} \tag{59}$$

2. $$\vec{z}_2 = z_1^{k_1} + \mu_2(\vec{a}_3 - \vec{z}_1^{k_1}) \tag{60}$$

3. for $\mu_2 = \mu_2^{k_2}$ (from Eq. (60)), we get:

$$\vec{z}_2^{k_2} = z_1^{k_1} + \mu_2^{k_2}(\vec{a}_3 - \vec{z}_1^{k_1}) \tag{61}$$

Taking into account Eqs. (59)–(61), write the expressions of the vectors $\overrightarrow{BC}$, $\overrightarrow{BA}$ and get:

$$\overrightarrow{BA} = \vec{z}_2^{k_2} - \vec{z}_3(\mu_3) = \vec{z}_2^{k_2} - [\vec{z}_2^{k_2} + \mu_3(\vec{a}_4 - \vec{z}_2^{k_2})] = \\ = \mu_3(\vec{a}_4 - \vec{z}_2^{k_2}) \tag{62}$$

$$\overrightarrow{BC} = \vec{z}_2(\mu_2) - \vec{z}_3(\mu_3) = (1 - \mu_2)\vec{z}_1^{k_1} + \mu_2 \vec{a}_3 - [\vec{z}_2^{k_2} +$$

$$\mu_3(\vec{a}_4 - \vec{z}_2^{k_2})] = \vec{z}_2(\mu_2) - \vec{z}_2^{k_2} - \mu_3(\vec{a}_4 - \vec{z}_2^{k_2})$$

or, allowing for Eqs. (59) and (61), get:

$$\overrightarrow{BC} = (\mu_2 - \mu_2^{k_2})(\vec{a}_3 - \vec{z}_1^{k_1}) - \mu_3[\vec{a}_4 - z_1^{k_1} - \mu_2^{k_2}(\vec{a}_3 - \vec{z}_1^{k_1})] \quad (63)$$

Substituting Eqs. (62) and (63) in vector product Eq. (58), we get:

$$\mu_3(\vec{a}_4 - \vec{z}_2^{k_2})\{(\mu_2 - \mu_2^{k_2})(\vec{a}_3 - \vec{z}_1^{k_1}) - \\ - \mu_3[\vec{a}_4 - \mu_2^{k_2}(\vec{a}_3 - \vec{z}_1^{k_1}) - \vec{z}_1^{k_1}]\} = 0 \quad (64)$$

From Eq. (64) it is seen that one of the values $\mu_3 = 0$. This means that the parameter μ_3 begins from the end of the second piecewise-linear straight line.

For an arbitrary value of the parameter μ_3, we get the following relation between the parameters:

$$\mu_3 = (\mu_2 - \mu_2^{k_2}) \frac{(\vec{a}_3 - \vec{z}_1^{k_1})(\vec{a}_4 - \vec{z}_2^{k_2})}{(\vec{a}_4 - \vec{z}_2^{k_2})^2},$$

for

$$\mu_2 \geq \mu_2^{k_2}, \ \mu_1 \geq \mu_1^{k_2} \quad (65)$$

Thus, by Eq. (65) we established mathematical relation of an arbitrary parameter μ_3 and parameter μ_2 that corresponds to the point of the second piecewise-linear straight line beginning from $\mu_2 = \mu_2^{k_2}$. On the other hand, by Eq. (34) the parameter μ_2 is connected with the parameter μ_1. If we substitute the value of μ_2 definable by Eq. (34), in Eq. (65), we get a catenary dependence of the parameter μ_3 on μ_1.

Now we can give Eq. (65) the following coordinate form:

$$\mu_3=(\mu_2-\mu_2^{k_2})\frac{\sum_{m=1}^{M}(a_{3m}-z_{1m}^{k_1})(a_{4m}-z_{2m}^{k_2})}{\sum_{m=1}^{M}(a_{4m}-z_{2m}^{k_2})^2}, \text{ for } \mu_2 \geqslant \mu_2^{k_2} \tag{66}$$

By similar calculations we can follow the method for constructing a vector equation for the fourth piecewise-linear straight line. According to the above-indicated method, set up the dependence of the points of the fourth piecewise-linear straight line on difference of economic process rate holding between the third and fourth piecewise-linear straight lines, in the following form:

$$\vec{z}_4=\vec{z}_1\{1+A\,[(1+\omega_2(\lambda_2^{k_2},\alpha_{1,2})+ +\omega_3(\lambda_3^{k_3},\alpha_{2,3})+\omega_4(\lambda_4,\alpha_{3,4})]\} \tag{67}$$

Here,

$$\omega_2(\lambda_2^{k_2},\alpha_{1,2})=\lambda_2^{k_2}\cos\alpha_{1,2}=$$

$$=\frac{\mu_2^{k_2}}{\mu_1^{k_1}-\mu_1}\cdot\frac{|\vec{z}_1-\vec{z}_1^{k_1}||\vec{a}_3-\vec{z}_1^{k_1}|}{\vec{z}_1(\vec{z}_1-\vec{z}_1^{k_1})}\cdot\frac{\vec{z}_1(\vec{z}_1^{k_1}-\vec{a}_1)}{|\vec{a}_2-\vec{a}_1||\vec{z}_1^{k_1}-\vec{a}_1|}\cos\alpha_{1,2}$$

$$\omega_3(\lambda_3^{k_3},\alpha_{2,3})=\lambda_3^{k_3}\cos\alpha_{2,3}=$$

$$=\frac{\mu_3^{k_3}}{\mu_1^{k_1}-\mu_1}\cdot\frac{|\vec{z}_2-\vec{z}_2^{k_2}||\vec{a}_4-\vec{z}_2^{k_2}|}{\vec{z}_1(\vec{z}_2-\vec{z}_2^{k_2})}\cdot\frac{\vec{z}_1(\vec{z}_1^{k_1}-\vec{a}_1)}{|\vec{a}_2-\vec{a}_1||\vec{z}_1^{k_1}-\vec{a}_1|}\cos\alpha_{2,3}$$

$$\omega_4(\lambda_4,\alpha_{3,4})=\lambda_4\cos\alpha_{3,4}=$$

$$= \frac{\mu_4}{\mu_1^{k_1} - \mu_1} \cdot \frac{\left|\vec{z}_3 - \vec{z}_3^{k_3}\right|\left|\vec{a}_5 - \vec{z}_3^{k_3}\right|}{\vec{z}_1(\vec{z}_3 - \vec{z}_3^{k_3})} \cdot \frac{\vec{z}_1(\vec{z}_1^{k_1} - \vec{a}_1)}{\left|\vec{a}_2 - \vec{a}_1\right|\left|\vec{z}_1^{k_1} - \vec{a}_1\right|} \cos\alpha_{3,4} \quad (68)$$

$$\mu_4 = (\mu_3 - \mu_3^{k_3}) \frac{(\vec{a}_4 - \vec{z}_2^{k_2})(\vec{a}_5 - \vec{z}_3^{k_3})}{(\vec{a}_5 - \vec{z}_3^{k_3})^2}, \text{ for } \quad \mu_3 > \mu_3^{k_3} \quad (69)$$

By the recurrent method it is easy to get in finite-dimensional Euclidean space the dependence of any n-th piecewise-linear vector equation $\vec{z}_n$ on the first piecewise-linear function $\vec{z}_1$ and all spatial type influence function of unaccounted parameters $\omega_n(\lambda_n^{k_n}, \alpha_{n-1,n})$ influencing on the preceding interval of economic event, in the form (Fig. 6):

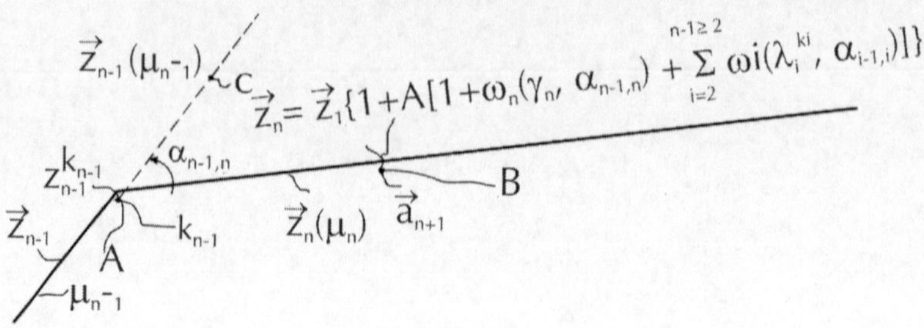

FIGURE 6 The scheme of construction of the n-th piecewise-linear vector equation in finite-dimensional vector space R_m.

$$\vec{z}_n = \vec{z}_1 \left\{ 1 + A \left[1 + \omega_n(\lambda_n, \alpha_{n-1,n}) + \sum_{i=2}^{n-1\geq 2} \omega_i(\lambda_i^{k_i}, \alpha_{i-1,i}) \right] \right\} \quad (70)$$

Here

$$\omega_i(\lambda_i^{k_i}, \alpha_{i-1,i}) = \lambda_i^{k_i} \cos\alpha_{i-1,i} =$$

$$= \frac{\mu_i^{k_i}}{\mu_1^{k_1} - \mu_1} \cdot \frac{\left|\vec{z}_{i-1} - \vec{z}_{i-1}^{k_{i-1}}\right|\left|\vec{a}_{i+1} - \vec{z}_{i-1}^{k_{i-1}}\right|}{\vec{z}_1(\vec{z}_{i-1} - \vec{z}_{i-1}^{k_{i-1}})} \cdot \frac{\vec{z}_1(\vec{z}_1^{k_1} - \vec{a}_1)}{\left|\vec{a}_2 - \vec{a}_1\right|\left|\vec{z}_1^{k_1} - \vec{a}_1\right|} \cos\alpha_{i-1,i} \quad (71)$$

$$\mu_i = (\mu_{i-1} - \mu_{i-1}^{k_{i-1}}) \frac{(\vec{a}_i - \vec{z}_{i-2}^{k_{i-2}})(\vec{a}_{i+1} - \vec{z}_{i-1}^{k_{i-1}})}{(\vec{a}_{i+1} - \vec{z}_{i-1}^{k_{i-1}})^2}, \text{ for } \mu_{i-1} \geqslant \mu_{i-1}^{k_{i-1}} \tag{72}$$

$$\omega_n(\lambda_n, \alpha_{n-1,n}) = \lambda_n cos\alpha_{n-1,n} =$$

$$= \frac{\mu_n}{\mu_1^{k_1} - \mu_1} \cdot \frac{|\vec{z}_{n-1} - \vec{z}_{n-1}^{k_{n-1}}||\vec{a}_{n+1} - \vec{z}_{n-1}^{k_{n-1}}|}{\vec{z}_1(\vec{z}_{n-1} - \vec{z}_{n-1}^{k_{n-1}})} \frac{\vec{z}_1(\vec{z}_1^{k_1} - \vec{a}_1)}{|\vec{a}_2 - \vec{a}_1||\vec{z}_1^{k_1} - \vec{a}_1|} cos\alpha_{n-1,n} \tag{73}$$

$$A = (\mu_1^{k_1} - \mu_1) \frac{|\vec{a}_2 - \vec{a}_1||\vec{z}_1^{k_1} - \vec{a}_1|}{\vec{z}_1(\vec{z}_1^{k_1} - \vec{a}_1)} \tag{74}$$

$$\mu_n = (\mu_{n-1} - \mu_{n-1}^{k_{n-1}}) \frac{(\vec{a}_n - \vec{z}_{n-2}^{k_{n-2}})(\vec{a}_{n+1} - \vec{z}_{n-1}^{k_{n-1}})}{(\vec{a}_{n+1} - \vec{z}_{n-1}^{k_{n-1}})^2}, \quad \mu_{n-1} \geqslant \mu_{n-1}^{k_{n-1}} \tag{75}$$

The value of $\cos\alpha_{n-1,n}$ between the (n–1)-th and the n-th piecewise linear vector equations is determined according to Fig. 6. in the form:

$$cos\alpha_{n-1,n} = \frac{(\vec{z}_{n-1} - \vec{z}_{n-1}^{k_{n-1}})(\vec{a}_{n+1} - \vec{z}_{n-1}^{k_{n-1}})}{|\vec{z}_{n-1} - \vec{z}_{n-1}^{k_{n-1}}||\vec{a}_{n+1} - \vec{z}_{n-1}^{k_{n-1}}|} \tag{76}$$

2.3 ON A CRITERION OF MULTIVARIANT PREDICTION OF ECONOMICAL PROCESS AND ITS CONTROL AT UNCERTAINTY CONDITIONS IN FINITE-DIMENSIONAL VECTOR-SPACE

Difficulty of economic process prediction and its control at uncertainty conditions is in multivariant character of the approach to the solution of the given problem.

In addition to this we should note that nonavailability of precise definition of the notion “uncertainty” in economic processes, incomplete classification of display of this phenomenon, and also nonavailability of its precise and clear mathematical representation places the finding of the solution of problems of prediction of economic process and this control to the higher level by its complexity.

Many-dimensionality and spatial inhomogeneity of the occurring economic process, time changeability of multifactor economic indices and also their change velocity give additional complexity and uncertainty.

Another complexity of the problem is connected with reliable construction of such a predicting vector equation in the consequent small volume $\Delta V_{n+1}(x_1, x_2, ... x_m)$ of finite-dimensional vector-space that could sufficiently reflect the state of economic process in the subsequent step. In other words, now by means of the given statistical points [illegible]ing certain economic process in the preceding volume $V = \sum_{N=1}^{N} \Delta V_N(x_1, x_2, ... x_m)$ of finite-dimensional vector space R_m one can construct a predicting vector equation $\vec{Z}_{n+1}(x_1, x_2, ... x_m)$ in the subsequent small volume $\Delta V_{n+1}(x_1, x_2, ... x_m)$ of finite-dimensional vector space.

The goal of our investigation is to formulate the notion of uncertainty for one class of economical processes and also to find mathematical representation of the predicting function $\vec{Z}_{n+1}(x_1, x_2, ... x_m)$ for the given class of processes depending on so-called unaccounted factors functions. In connection with what has been said, below we suggest a method for constructing a predicting vector equation $\vec{Z}_{n+1}(x_1, x_2, ... x_m)$ in the subsequent small volume $\Delta V_{n+1}(x_1, x_2, ... x_m)$ of finite-dimensional vector space [11, 14, 23–29].

In Section 2.2, the postulate spatial-time certainty of economic process at uncertainty conditions in finite-dimensional vector space" was suggested, the notion of piecewise-linear homogeneity of the occurring economic process at uncertainty conditions was introduced, and also a so called the unaccounted parameters influence function $\omega_n(\lambda_n^{k_n}, \alpha_{n-1\,n})$ influencing on the preceding volume $V = \sum_{N=1}^{N} \Delta V_N$ of economic process was suggested. On this basis, it was suggested the dependence of the n-th piecewise-linear function $\vec{z}_n$ on the first piecewise-linear function $\vec{z}_1$ and all spatial type unaccounted parameters influence function $\omega_n(\lambda_n^{k_n}, \alpha_{n-1\,n})$ influencing on the preceding interval of economic process, in the form Eqs. (70)–(75):

$$\vec{z}_n = \vec{z}_1 \left\{ 1 + A \left[1 + \omega_n(\lambda_n, \alpha_{n-1,n}) + \sum_{i=2}^{n-1 \geq 2} \omega_i(\lambda_i^{k_i}, \alpha_{i-1,i}) \right] \right\} \tag{77}$$

where

$$\omega_i(\lambda_i^{k_i}, \alpha_{i-1,i}) = \lambda_i^{k_i} cos\alpha_{i-1,i} =$$

$$= \frac{\mu_i^{k_i}}{\mu_1^{k_1} - \mu_1} \cdot \frac{\left|\vec{z}_{i-1} - \vec{z}_{i-1}^{k_{i-1}}\right| \left|\vec{a}_{i+1} - \vec{z}_{i-1}^{k_{i-1}}\right|}{\vec{z}_1(\vec{z}_{i-1} - \vec{z}_{i-1}^{k_{i-1}})} \cdot \frac{\vec{z}_1(\vec{z}_1^{k_1} - \vec{a}_1)}{\left|\vec{a}_2 - \vec{a}_1\right| \left|\vec{z}_1^{k_1} - \vec{a}_1\right|} cos\alpha_{i-1,i} \tag{78}$$

are unaccounted parameters influence functions influencing on the preceding ΔV_1, ΔV_2,..ΔV_i small volumes of economic process;

$$\mu_i = (\mu_{i-1} - \mu_{i-1}^{k_{i-1}}) \frac{(\vec{a}_i - \vec{z}_{i-2}^{k_{i-2}})(\vec{a}_{i+1} - \vec{z}_{i-1}^{k_{i-1}})}{(\vec{a}_{i+1} - \vec{z}_{i-1}^{k_{i-1}})^2}, \text{ for } \mu_{i-1} > \mu_{i-1}^{k_{i-1}} \tag{79}$$

are arbitrary parameters referred to the i-th piecewise-linear straight line. And the parameters μ_i are connected with the parameter μ_{i-1} referred to the (i-1)-th piecewise-linear straight line, in the form Eq. (79);

$$A = (\mu_1^{k_1} - \mu_1) \frac{\left|\vec{a}_2 - \vec{a}_1\right| \left|\vec{z}_1^{k_1} - \vec{a}_1\right|}{\vec{z}_1(\vec{z}_1^{k_1} - \vec{a}_1)} \tag{80}$$

is a constant quantity;

$$\omega_n(\lambda_n, \alpha_{n-1,n}) = \lambda_n cos\alpha_{n-1,n} =$$

$$= \frac{\mu_n}{\mu_1^{k_1} - \mu_1} \cdot \frac{\left|\vec{z}_{n-1} - \vec{z}_{n-1}^{k_{n-1}}\right| \left|\vec{a}_{n+1} - \vec{z}_{n-1}^{k_{n-1}}\right|}{\vec{z}_1(\vec{z}_{n-1} - \vec{z}_{n-1}^{k_{n-1}})} \cdot \frac{\vec{z}_1(\vec{z}_1^{k_1} - \vec{a}_1)}{\left|\vec{a}_2 - \vec{a}_1\right| \left|\vec{z}_1^{k_1} - \vec{a}_1\right|} cos\alpha_{n-1,n} \tag{81}$$

is the expression of the unaccounted parameters influence function that influences on subsequent small volume ΔV_N of finite-dimensional vector space. And the parameter μ_n referred to the n- piecewise-linear straight line is of the form:

$$\mu_n = (\mu_{n-1} - \mu_{n-1}^{k_{n-1}}) \frac{(\vec{a}_n - \vec{z}_{n-2}^{k_{n-2}})(\vec{a}_{n+1} - \vec{z}_{n-1}^{k_{n-1}})}{(\vec{a}_{n+1} - \vec{z}_{n-1}^{k_{n-1}})^2}, \quad \mu_{n-1} \geq \mu_{n-1}^{k_{n-1}} \tag{82}$$

Here the parameter μ_n is connected with the parameter μ_{n-1} of the preceding (n-1)-th piecewise-linear vector equation of the straightline in the form Eq. (82).

Thus, in finite-dimensional vector space, the system of statistical points (vectors) is represented in the vector form in the form of N piecewise-linear straight lines depending on the vector function of the first piecewise-linear straight-line $\vec{z}_1 = \lambda_1 \vec{a}_1 + \mu_1 \vec{a}_2$, and also on the unaccounted parameters influence function $\omega_n(\lambda_n, \alpha_{n-1\,n})$ in all the investigated preceding volume of finite-dimensional vector space R_m.

Now let's try to solve a problem on prediction of economic process and its control at uncertainty conditions in finite-dimensional vector space.

For that at first we note some peculiarities of the introduced "unaccounted parameters influence function" $\omega_n(\lambda_n, \alpha_{n-1\,n})$. The unaccounted parameters influence functions $\omega_n(\lambda_n, \alpha_{n-1\,n})$ are integral characteristics of influencing external factors occurring in environment that are not a priori situated in functional chain of sequence of the structured model but render very strong functional influence both on the function and on the results of prediction quantities Eq. (77). It is impossible to fix such a cause by statistical means. This means that the investigated this or other economic process in finite dimensional vector space directly or obliquely is connected with many dimensionality and spatial inhomogenlity of the occurring economic process, with time changeability of multifactor economic indices, vector and their change velocity. This in its turn is connected with the fact that the used statistical data of economic process in finite-dimensional vector space are of inhomogeneous in coordinates and time unstationary events character.

For systematizing the investigated economic process we distinguish a class that possesses precisely expressed form of uncertainty. We assume the given unaccounted factors functions $\omega_n(\lambda_n, \alpha_{n-1\,n})$ hold on all the preceding interval of finite-dimensional vector space, the uncertainty character of this class of economic process.

In such a statement, the problem on prediction of economic event on the subsequent small volume ΔV_{N+1} of finite-dimensional vector space will be directly connected in the first turn with the enumerated invisible external facts fixed on the earlier stages and their combinations, i.e., the functions $\omega_n(\lambda_n, \alpha_{n-1\,n})$ that earlier hold in the preceding small volumes $\Delta V_1, \Delta V_2, ...\Delta V_N$ of finite-dimensional vector space. Therefore, by studying the problem on prediction of economic process on subsequent small volume ΔV_{N+1} it is necessary to be ready to possible influence of such factors.

In connection with such a statement of the problem, let's investigate behavior of economic process in subsequent small volume ΔV_{N+1} finite-dimensional vector space under the action of the desired unaccounted parameters function $\omega_n(\lambda_n, \alpha_{n-1\,n})$ that was earlier fixed by us in preceding small volumes ΔV_n of finite-dimensional vector space, i.e., $\omega_2(\lambda_2, \alpha_{1\,2})$, $\omega_3(\lambda_3, \alpha_{2\,3})$, ..., $\omega_N(\lambda_N, \alpha_{N-1\,N})$.

In connection with what has been said, the problem on prediction of economic process and its control in finite-dimensional vector space may be solved by means of the introduced unaccounted parameters influence function $\omega_n(\lambda_n, \alpha_{n-1,n})$ in the following way.

Construct the (N+1)-th vector equation of piecewise-linear straight line $\vec{z}_{N+1} = \vec{z}_N^{k_N} + \mu_{N+1}(\vec{a}_{N+2} - \vec{z}_N^{k_N})$ depending on the vector equation of the first piecewise-linear straight line $\vec{z}_1$ and the desired unaccounted parameter influence function $\omega_\beta(\lambda_\beta, \alpha_{\beta-1,\beta}$ that we have seen in one of the preceding small volumes $\Delta V_1, \Delta V_2, ...\Delta V_N$ of finite-dimensional vector space. For that in Eqs. (77)–(82) we change the index n by $(N+1)$ and get:

$$\vec{z}_{N+1} = \vec{z}_1 \left\{ 1 + A \left[1 + \sum_{i=2}^{N} \omega_i(\lambda_i^{k_i}, \alpha_{i-1,i}) + \omega_{N+1}(\lambda_{N+1}, \alpha_{N,N+1}) \right] \right\} \tag{83}$$

Here

$$\omega_i(\lambda_i^{k_i}, \alpha_{i-1,i}) = \lambda_i^{k_i} cos\alpha_{i-1,i} =$$
$$= \frac{\mu_i^{k_i}}{\mu_1^{k_1} - \mu_1} \cdot \frac{\left|\vec{z}_{i-1} - \vec{z}_{i-1}^{k_{i-1}}\right| \left|\vec{a}_{i+1} - \vec{z}_{i-1}^{k_{i-1}}\right|}{\vec{z}_1(\vec{z}_{i-1} - \vec{z}_{i-1}^{k_{i-1}})} \cdot \frac{\vec{z}_1(\vec{z}_1^{k_1} - \vec{a}_1)}{\left|\vec{a}_2 - \vec{a}_1\right| \left|\vec{z}_1^{k_1} - \vec{a}_1\right|} cos\alpha_{i-1,i} \tag{84}$$

$$\mu_i = (\mu_{i-1} - \mu_{i-1}^{k_{i-1}}) \frac{(\vec{a}_i - \vec{z}_{i-2}^{k_{i-2}})(\vec{a}_{i+1} - \vec{z}_{i-1}^{k_{i-1}})}{(\vec{a}_{i+1} - \vec{z}_{i-1}^{k_{i-1}})^2}, \qquad \mu_{i-1} > \mu_{i-1}^{k_{i-1}} \tag{85}$$

$$A = (\mu_1^{k_1} - \mu_1) \frac{|\vec{a}_2 - \vec{a}_1| |\vec{z}_1^{k_1} - \vec{a}_1|}{\vec{z}_1 (\vec{z}_1^{k_1} - \vec{a}_1)} \tag{86}$$

$$\omega_{N+1}(\lambda_{N+1}, \alpha_{N,N+1}) = \lambda_{N+1} \cos\alpha_{N,N+1} =$$

$$= \frac{\mu_{N+1}}{\mu_1^{k_1} - \mu_1} \cdot \frac{|\vec{z}_N - \vec{z}_N^{k_N}| |\vec{a}_{N+2} - \vec{z}_N^{k_N}|}{\vec{z}_1(\vec{z}_N - \vec{z}_N^{k_N})} \cdot \frac{\vec{z}_1(\vec{z}_1^{k_1} - \vec{a}_1)}{|\vec{a}_2 - \vec{a}_1| |\vec{z}_1^{k_1} - \vec{a}_1|} \cos\alpha_{N,N+1} \tag{87}$$

$$\mu_{N+1} = (\mu_N - \mu_N^{k_N}) \frac{(\vec{a}_{N+1} - \vec{z}_{N-1}^{k_{N-1}})(\vec{a}_{N+2} - \vec{z}_N^{k_N})}{(\vec{a}_{N+2} - \vec{z}_N^{k_N})^2}, \mu_N > \mu_N^{k_N} \tag{88}$$

For the behavior of economic process on the subsequent small volume ΔV_{N+1} of finite-dimensional vector space to be as in one of the desired preceding ones in small volume ΔV_β it is necessary that the vector equations of piecewise-linear straight lines $\vec{z}_{N+1}$ and $\vec{z}_\beta$ to be situated in one of the planes of these vectors and to be parallel to one another, i.e.

$$\vec{z}_{N+1} = C\vec{z}_\beta \tag{89}$$

In connection with what has been said, to ΔV_{N+1} finite-dimensional space there should be chosen such a vector-point $\vec{a}_{N+2}$ that the piecewise-linear straight lines $\vec{z}_{N+1} = (\vec{a}_{N+2} - \vec{z}_N^{k_N})$ and $\vec{z}_\beta = (\vec{a}_{\beta+1} - \vec{z}_{\beta-1}^{k_{\beta-1}})$ could be situated in the same plane of these vectors and at the same time be parallel to each other (Fig. 7).

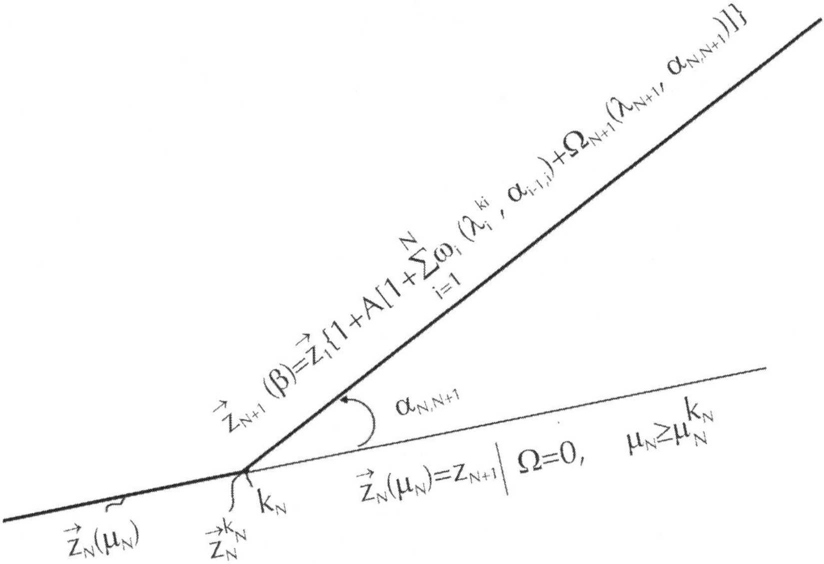

FIGURE 7 The scheme of construction of prediction function of economic process $\vec{z}_{N+1}(\beta)$ at uncertainty conditions in finite-dimensional vector space R_m. Prediction function $\vec{z}_{N+1}(\beta)$ will lie in the same plane with one of the desired preceding β-th piecewise-linear straight line and will be parallel to it.

In other words, they should satisfy the following parallelism condition:

$$(\vec{a}_{N+2} - \vec{z}_N^{k_N}) = C(\vec{a}_{\beta+1} - \vec{z}_{\beta-1}^{k_{\beta-1}}) \tag{90}$$

Here

$$\vec{a}_{N+2} = \sum_{m=1}^{M} a_{N+2,m}\vec{i}_m, \quad \vec{a}_{\beta+1} = \sum_{m=1}^{M} a_{\beta+1,m}\vec{i}_m,$$

$$\vec{z}_N^{k_N} = \sum_{m=1}^{M} z_{N,m}^{k_N}\vec{i}_m, \quad \vec{z}_{\beta-1}^{k_{\beta-1}} = \sum_{m=1}^{M} z_{\beta-1,m}^{k_{\beta-1}}\vec{i}_m$$

Excluding in Eq. (90) the parameter <u>C</u>, we get:

$$\frac{a_{N+2,1}-z_{N,1}^{k_N}}{a_{\beta+1,1}-z_{\beta-1,1}^{k_{\beta-1}}}=\frac{a_{N+2,2}-z_{N,2}^{k_N}}{a_{\beta+1,2}-z_{\beta-1,2}^{k_{\beta-1}}}=.....=\frac{a_{N+2,M}-z_{N,M}^{k_N}}{a_{\beta+1,M}-z_{\beta-1,M}^{k_{\beta-1}}} \tag{91}$$

It is easy to define from system Eq. (91) the coefficients of the vector $\vec{a}_{N+2}$:

$$a_{N+2,2}=z_{N,2}^{k_N}+\frac{a_{\beta+1,2}-z_{\beta-1,2}^{k_{\beta-1}}}{a_{\beta+1,1}-z_{\beta-1,1}^{k_{\beta-1}}}(a_{N+2,1}-z_{N,1}^{k_N})$$

$$a_{N+2,3}=z_{N,3}^{k_N}+\frac{a_{\beta+1,3}-z_{\beta-1,3}^{k_{\beta-1}}}{a_{\beta+1,1}-z_{\beta-1,1}}(a_{N+2,1}-z_{N,1}^{k_N})$$

$$a_{N+2,M}=z_{N,M}^{k_N}+\frac{a_{\beta+1,M}-z_{\beta-1,M}^{k_{\beta-1}}}{a_{\beta+1,M-1}-z_{\beta-1,M-1}^{k_{\beta-1}}}(a_{N+2,M-1}-z_{N,M-1}^{k_N}) \tag{92}$$

In this case, the vector $\vec{a}_{N+2}$ will have the following final form:

$$\vec{a}_{N+2}=a_{N+2,1}\vec{i}_1+a_{N+2,2}\vec{i}_2+a_{N+2,3}\vec{i}_3+......+a_{N+2,M}\vec{i}_M \tag{93}$$

As the coordinates of the point (of the vector) $\vec{a}_{N+2}$ now are determined by means of the piecewise-linear vector $\vec{z}_\beta=\vec{a}_{\beta+1}-\vec{z}_{\beta-1}^{k_{\beta-1}}$ taken from one of the preceding stage of economic process, it is appropriate to denote them in the form $\vec{a}_{N+2}(\beta)$. This will show that the coordinates of the point $\vec{a}_{N+2}$ (3) were determined by means of piecewise-linear straight line $\vec{z}_\beta$. In this case it is appropriate to represent Eq. (93) in the following compact form:

$$\vec{a}_{N+2}(\beta)=\sum_{m=1}^{M}a_{N+2,m}(\beta)\vec{i}_m \tag{94}$$

Now, in the system of Eqs. (83)–(88), instead of the vector $\vec{a}_{N+1}$ we substitute the value of the vector $\vec{a}_{N+2}(\beta)$, and also instead of $\omega_{N+1}(\lambda_{N+1},\alpha_{N,N+1})$ introduce the denotation of the so-called predicting influence function with regard to unaccounted parameters $\Omega_{N+1}(\lambda_{N+1}, \alpha_{N,N+1})$. In this case the prediction function of the economic process $\vec{Z}_{N+1}(\beta)$ with regard to influence of prediction function of unaccounted parameters $\Omega_{N+1}(\lambda_{N+1}, \alpha_{N,N+1})$ will take the following form:

$$\vec{Z}_{N+1}(\beta)=\vec{z}_1\left\{1+A\left[1+\sum_{i=2}^{N}\omega_i(\lambda_i^{k_i}, \alpha_{i-1,i})+\Omega_{N+1}(\lambda_{N+1}, \alpha_{N,N+1})\right]\right\} \quad (95)$$

Here

$$\omega_i(\lambda_i^{k_i}, \alpha_{i-1,i}) = \lambda_i^{k_i} cos\alpha_{i-1,i} =$$

$$= \frac{\mu_i^{k_i}}{\mu_1^{k_1} - \mu_1} \cdot \frac{\left|\vec{z}_{i-1} - \vec{z}_{i-1}^{k_{i-1}}\right|\left|\vec{a}_{i+1} - \vec{z}_{i-1}^{k_{i-1}}\right|}{\vec{z}_1(\vec{z}_{i-1} - \vec{z}_{i-1}^{k_{i-1}})} \cdot \frac{\vec{z}_1(\vec{z}_1^{k_1} - \vec{a}_1)}{\left|\vec{a}_2 - \vec{a}_1\right|\left|\vec{z}_1^{k_1} - \vec{a}_1\right|} cos\alpha_{i-1,i} \quad (96)$$

$$\mu_i=(\mu_{i-1} - \mu_{i-1}^{k_{i-1}})\frac{(\vec{a}_i - \vec{z}_{i-2}^{k_{i-2}})(\vec{a}_{i+1} - \vec{z}_{i-1}^{k_{i-1}})}{(\vec{a}_{i+1} - \vec{z}_{i-1}^{k_{i-1}})^2}, \quad \mu_{i-1} > \mu_{i-1}^{k_{i-1}} \quad (97)$$

$$A = (\mu_1^{k_1} - \mu_1)\frac{\left|\vec{a}_2 - \vec{a}_1\right|\left|\vec{z}_1^{k_1} - \vec{a}_1\right|}{\vec{z}_1(\vec{z}_1^{k_1} - \vec{a}_1)} \quad (98)$$

and the prediction function of influence of unaccounted parameters $\Omega_{N+1}(\lambda_{N+1}, \alpha_{N,N+1})$ will take the form:

$$\Omega_{N+1}(\lambda_{N+1}, \alpha_{N,N+1}) = \lambda_{N+1} cos\alpha_{N,N+1} \quad (99)$$

$$\lambda_{N+1} = \frac{\mu_{N+1}}{\mu_1^{k_1} - \mu_1} \cdot \frac{\left|\vec{z}_N - \vec{z}_N^{k_N}\right|\left|\vec{a}_{N+2}(\beta) - \vec{z}_N^{k_N}\right|}{\vec{z}_1(\vec{z}_N - \vec{z}_N^{k_N})} \cdot \frac{\vec{z}_1(\vec{z}_1^{k_1} - \vec{a}_1)}{\left|\vec{a}_2 - \vec{a}_1\right|\left|\vec{z}_1^{k_1} - \vec{a}_1\right|} \quad (100)$$

$$\mu_{N+1}=(\mu_N - \mu_N^{k_N})\frac{(\vec{a}_{N+1} - \vec{z}_{N-1}^{k_{N-1}})(\vec{a}_{N+2}(\beta) - \vec{z}_N^{k_N})}{(\vec{a}_{N+2}(\beta) - \vec{z}_N^{k_N})^2}, \quad \mu_N > \mu_N^{k_N} \quad (101)$$

Here the vector $\vec{a}_{N+2}(\beta)$ is determined by Eq. (94).

Note the following points. It is seen from Eq. (101) that for $\mu_N = \mu_N^{k_N}$ the value of the parameter $\mu_{N+1} = 0$. By this fact from Eq. (99) it will follow that the value of the predicting function of influence of unaccounted parameters $\Omega_{N+1}(\lambda_{N+1}, \alpha_{N,N+1})$ will equal:

$$\Omega_{N+1}(\lambda_{N+1}, \alpha_{N,N+1}) = 0 \text{ for } \mu_{N+1} = 0$$

$$\Omega_{N+1}(\lambda_{N+1}, \alpha_{N,N+1}) \neq 0 \text{ for } \mu_{N+1} > 0 \tag{102}$$

This will mean that the initial point from which the (N+1)-th vector equation of the prediction function of economic process $\vec{Z}_{N+1}(\beta)$ will enanimate, will coincide with the final point of the n-th vector equation of piecewise-linear straight line $\vec{z}_N$ and equal:

$$Z_{N+1} = \vec{z}_1 \left\{ 1 + A \left[1 + \sum_{i=2}^{N} \omega_i(\lambda_i^{k_i}, \alpha_{i-1,i}) \right] \right\}, \quad \text{for } \mu_{N+1} = 0 \tag{103}$$

For any other values of the parameter $\mu_{N+1} \neq 0$ the points of the $(N+1)$-th vector equation will be determined by Eq. (95).

It is seen from Eq. (99) that $\lambda_{N+1} = 0$ and $\Omega_{N+1}(\lambda_{N+1} = 0; \alpha_{N,N+1}) = 0$ will follow $cos\alpha_{N,N+1} = 0$ and $\mu_{N+1} \neq 0$. This will correspond to the case when the influence of external unaccounted factors on subsequent small volume ΔV_{N+1} are as in the preceding small volume ΔV_N of finite-dimensional vector space. In this case it suffices to continue the preceding vector equation $\vec{z}_N$ to the desired point $\mu_{N+1} = \mu_{N+1}^* > \mu_N^{k_N}$ of subsequent small volume of finite-dimensional vector space. The value of the vector function $\vec{Z}_{N+1}(\mu_{N+1}^*) = \vec{z}_N(\mu_{N+1}^*; \lambda_N, \alpha_{N-1,N})$ at the point $\mu_{N+1} = \mu_{N+1}^*$ will be one of the desired prediction values of economic process in subsequent small volume ΔV_{N+1}. In this case, the value of the controlled parameter of unaccounted factors will be equal to zero, i.e.,

$$\Omega_{N+1}(\mu_{N+1} \neq 0; \lambda_{N+1} \neq 0; \cos\alpha_{N,N+1} = 0; \alpha_{N,N+1} = 0) = 0$$

For any other value of the parameter μ_{N+1}, taken on the interval $0\leq \mu_{N+1}\leq \mu^*_{N+1}$ and $\cos\alpha_{N,N+1}\neq 0$, the corresponding prediction function of unaccounted parameters will differ from zero, i.e., $\Omega_{N+1}(\lambda_{N+1},\alpha_{N,N+1})\neq 0$. Thus, choosing by desire the numerical values of unaccounted parameters function $\omega_\beta(\mu_{N+1};\lambda_\beta,\alpha_{\beta-1,\beta})=\Omega_{N+1}(\lambda^*_{N+1},\alpha_{N,N+1})$ corresponding to preceding small volumes $\Delta V_1,\Delta V_2,.....\Delta V_N$ and influencing by them beginning with the point $\mu_{N+1}=0$ to the desired point μ^*_{N+1}, we get numerical values of predicting economic event $\vec{Z}_{N+1}(\mu^*_{N+1};\lambda^*_{N+1},\alpha_{N,N+1})$ on subsequent step of the small volume ΔV_{N+1} (Fig. 8).

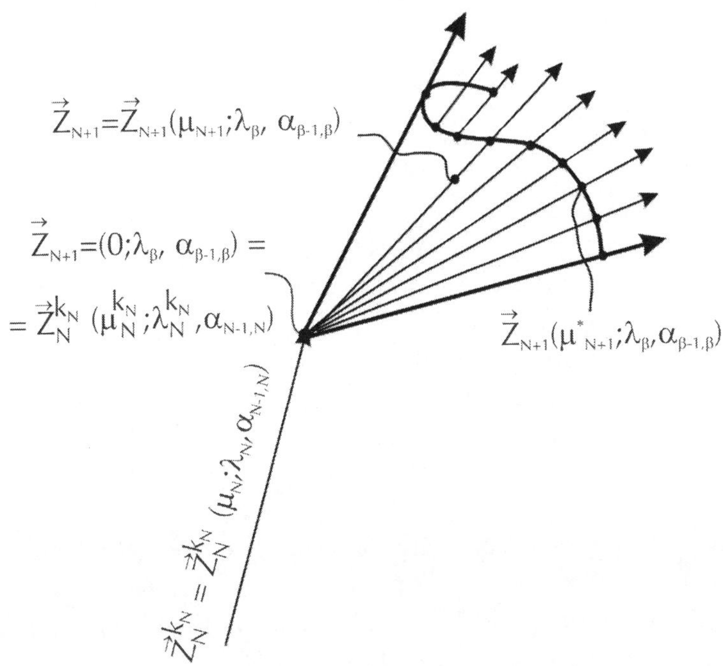

FIGURE 8 The graph of prediction of process and its control at uncertainty conditions in finite-dimensional vector space. It is represented in the form of hyperconic surface whose points, of directrix will form the line of economic process prediction.

Taking into account the fact that by desire we can choose the predicting influence function of unaccounted parameters $\Omega^*_{N+1}(\mu^*_{N+1};\lambda^*_{N+1},\alpha_{N,N+1})$, then this function will represent a predicting control function of unaccounted factors, and its appropriate function $\vec{Z}^*_{N+1}(\mu^*_{N,N+1};\lambda^*_{N+1},\alpha_{N,N+1})$ will be a control aim function of economic event in finite-dimensional vector space. Speaking about unaccounted parameters prediction function $\Omega_{N+1}(\mu_{N+1};\lambda_{N+1},\alpha_{N,N+1})$ we should understand their preliminarily calculated values in previous small volumes $\Delta V_1,\Delta V_2,\ldots..\Delta V_N$ of finite-dimensional vector space. Therefore, in Eq. (95) we used calculated ready values of the function $\Omega_{N+1}(\mu_{N+1};\lambda_{N+1},\alpha_{N,N+1})$.
Thus, influencing by the unaccounted parameters influence functions of the form $\Omega_{N+1}(\mu_{N+1};\lambda_{N+1},\alpha_{N,N+1})$ or by their combinations from the end of the vector equation of piecewise-linear straight line $\vec{z}^{k_N}_N(\mu^{k_N}_N;\ \lambda^{k_N}_N,\alpha_{N\text{-}1\ N})$ situated on the boundary of the small volume $\vec{Z}_{N+1}(\beta)=\vec{Z}_{N+1}(\mu_{N+1};\lambda_{N+1},\alpha_{N,N+1})$ there will originate the vectors ΔV_N and ΔV_{N+1}, lying on the subsequent small volume ΔV_{N+1}.
These vectors will represent the generators of hyperbolic surface of finite-dimensional vector space. The values of this series vector-functions for small values of the parameter $\mu_{N+1}=\mu^*_{N+1}$, i.e., $\vec{Z}_{N+1}(\mu^*_{N+1};\lambda_{N+1},\alpha_{N,N+1})$ will represent the points directrix of hyperconic surface of finite-dimensional vector space (Fig. 8). The series of the values of the points of directrix hyperconic surface will create a domain of change of predictable values of the function of $\vec{Z}^*_{N+1}(\mu^*_{N+1};\lambda^*_{N+1},\alpha_{N,N+1})$ in the further step in the small volume ΔV_{N+1}. This predictable function will have minimum and maximum of its values $[\vec{Z}^*_{N+1}(\mu^*_{N+1};\lambda^*_{N+1},\alpha_{N,N+1})]_{\min}$ and $[\vec{Z}^*_{N+1}(\mu^*_{N+1};\lambda^*_{N+1},\alpha_{N,N+1})]_{max}$.

Thus, the found domain of change of predictable function of economic process in the form $\vec{Z}_{N+1}(\mu^*_{N+1};\lambda^*_{N+1},\alpha_{N,N+1})$, or in other words, the points of directrix of hyperbolic surface will represent the domain of economic process control in finite-dimensional vector-space [30–37].

KEYWORDS

- **finite-dimensional vector space**
- **hyperbolic surface**
- **piecewise-linear vector-function**
- **unaccounted parameters influence function**

REFERENCES

1. Aliyev Azad, G. *On a dynamical model for investigating economic problems.* Proceedings of the IMM NAS of Azerbaijan, Baku, ISSN 0207-3188, Baku, **1998,** *9,* 195–203.
2. Aliyev Azad, G. *Ekonomi Meselelerin Cozumune Iliskin Matematik Dinamik Bir Model.* Dergi 'Bilgi ve Toplum,' Turkiye, Istanbul, Milli Yayin, 99-34-Y-0147, ISBN 975-478-120-5, **1999,** 83–106.
3. Aliyev Azad, G. *Development of dynamical model for economic process and its application in the field of industry of Azerbaijan.* Thesis for PHD; Baku, **2001,** 167.
4. Aliyev Azad, G. *Construction methods of economic-mathematical models, prediction and control of economic event in incomplete information conditions.* Proceedings of the II International scientific-practical conference of young scientists. Alma-aty, Kazakh National Technical University, ISBN 9965-585-50-4, **2002,** 159–162.
5. Aliyev Azad, G.; Ecer, F. *Tam olmayan bilqiler durumunda iktisadi matematik metodlar ve modeler,* Ed. NUI of Turkiye, ISBN 975-8062-1802, Nigde, 2004, 223.
6. Aliyev Azad, G. *Persepectives of development of the methods of modeling and prediction of economic processes.* Izv. NAS of Azerbaijan, Ser. of hummanitarian and social sciences (economics), Baku, **2006,** *2,* 253–260.
7. Aliyev Azad, G. *A construction methods for piecewise-linear economic-mathematical models with regard to uncertainty factor in finite-dimensional vector space.* Collection of papers of the Institute of Economics of NAS of Azerbaijan (Problems of national economics), ISBN-5-8066-1711-3, Baku, **2007,** *2,* 290–301.
8. Aliyev Azad, G. *Issues of forecasting and management of economic processes in view of the factor of uncertainty in finite-dimensional vector space.* PCI, The second international conference "Problem of cybernetics and informatics," ISBN 978-9952-434-09-5, Baku, **2008,** *3,* 137–140.
9. Aliyev Azad, G. *Bases of piecewise-linear economic-mathematical models with regard to unaccounted factors influence on a plane and multivariant prediction of economic event.* "Issues of economics science" "Sputnik," ISSN 1728-8878, Moscow, **2009,** *3,* 187–201.
10. Aliyev Azad, G. *Definition of optimal plan of equipment in oilgas industry by means of dynamical economic-mathematical programming.* Izvstiya NAS of Azerbaijan ser. human. and social sciences (economics). Baku**, 2009,** *4,* 37–42.
11. Aliyev Azad, G. *Theoretical aspects of the problem of analysis of risks of investigation projects in oilgas recovery industry.* Izv. of NASA Ser. of human and social sciences (economics), Baku, **2009,** *1,* 97–104.
12. Aliyev Azad, G. *On development of four-component piecewise-linear economic-mathematical model with regard to uncertainty factor in finite-dimensional vector space.* Izvestia NAS of Azerbayjan, (economics), Baku, **2007,** *3,* 96–102.
13. Aliyev Azad, G. *Certainty criterion of economic event in finite-dimensional vector space.* Vestnik Khabarovskogo KhGAP, Khabarovsk, **2008,** *3(36),* 26–31. (Russian)
14. Aliyev Azad, G. *Economic-mathematical methods and models in uncertainty conditions in finite-dimensional vector space.* Ed. NAS of Azerbaijan "Information technologies." ISBN 978-9952-434-10-1, Baku, **2009,** 220.

15. Aliyev Azad, G. *Piecewise-mathematical models with regard to uncertainty factor in finite-dimensional vector space*. "Economika, Informatika" Vestnik, U.M.O.: Moscow, **2008,** *3,* 34–38.
16. Aliyev Azad, G. *A model of external factors influence in the course of economic process.* Proceedings of the IV International conference "Economics and vital activity" State University Sumgait, **2002,** 172–173.
17. Aliyev Azad, G. *A model for defining the required profit and price for accumulation of the planned volume of proper financial means in fuzzy condition.* Journal "Knowledge" "Education" Society of Azerbaijan Republic, Social Sciences three–four, 1683–7649, Baku, **2005,** 27–31.
18. Aliyev Azad, G. *Search of decisions in fuzzy conditions*. Izv. Visshikh tekhnicheskikh uchebnikh zavedeniy Azerbayjan. ASOA, ISSN 1609–1620, Baku, **2006,** *5(45)*, 66–71.
19. Aliyev Azad, G. *On construction of conjugate vector in Euclidean space for economic-mathematical modeling.* Izvestia NASA Ser. of Humanitarian and social sciences (economics), Baku, **2007,** *2,* 242–246.
20. Aliyev Azad, G. *A construction method for piecewise-linear economic-mathematical models in uncertainty conditions in finite-dimensional vector space.* Collection of papers of the XVI International Conference "Mathematics, Economics, Education," Rostov-na-donu, **2008,** 131–132.
21. Aliyev Azad, G. *On construction of piecewise-linear economic-mathematical models in uncertainty conditions in finite-dimensional vector space*, Vestnik KhGAP, Khabarovsk, **2008,** *5(38)*, 34–41.
22. Aliyev Azad, G. *On a problem of economic process prediction and control in finite-dimensional vector space.* Vestnik Kh GAP, Khabarovsk, **2008,** *6(39),* 31–39.
23. Aliyev Azad, G. *Peculiarities of construction of dynamical modles with regard to unaccounted factors.* Reports of Republican scientific-practical conference "Strategic problems of Azerbaijan economics," ASOA, Baku, **2002,** 81–84.
24. Aliyev Azad, G. *Difficulties of construction of economic-mathematical models, prediction and control of economic event in incomplete information conditions.* "Urgent problems of economics of transitional period" (collection of scientific papers). ASOA publishing, Baku, **2002,** 318–323.
25. Aliyev Azad, G.; Ismailova L. G. *On one way of prediction and control of economic event in contemporary market conditions.* Reports of the Republican scientific practical conference "Strategical problems of economics of Azerbaijan," ASOA, Baku, **2002,** 87–90.
26. Aliyev Azad, G. *On a mathematical aspect of constructing a conjugate vector for defining the unaccounted factors influence function of economic process in finite-dimensional vector space,* "Urgent social-economic problems of oil, chemical and machine-building fields of economics" (collection of scientific papers of the faculty "International Economic relations and Management" of Azerbaijan. State Oil Academy) Baku, **2007,** 600–616.
27. Aliyev Azad, G.; Farzaliyev M. M. *Some methodological aspects of developments of economic-mathematical models.* ASOA Reports of Republishing Scientific-practical conference "Development of Sumgait in contemporary market conditions, problems and perspectives. SSU.; Sumgait, **2007,** 80–81.

28. Aliyev Azad, G. *On a criterion of prediction and control of economic process with regard to uncertainty factor in finite-dimensional vector space.* "Iqtisadiyyat ve Hayat," ISSN 0207–3021, Baku, **2008,** *5,* 49–56.
29. Aliyev Azad, G. *On a criterion of economic process certainty in finite-dimensional vector space.* "Economic, statistic and Informatical" Vestnik of, U. M. O.: Moscow, **2008,** *2,* 33–37.
30. Aliyev Azad, G. *On a principle of prediction and control of economic process with regard to uncertainty factor in one-dimensional vector space,* "Economic, Statistical and Informatical" Vestnik, U.M.O.: Moscow, **2008,** *4,* 27–32.
31. Aliyev Azad, G. *On a way for defining the unknown parameters function of inhomogeneous economic process in finite-dimensional vector space.* Scientific and pedagogical Izvestia of "Odlar Yurdu" (ser. Of phys. Math. Techn. I natural sci.), ISSN 1682-9123, Baku, **2007,** *18,* 88–97.
32. Aliyev Azad, G. *Construction of two-component piecewise-linear economic-mathematical model with regard to uncertainty factor in finite-dimensional vector space.* "Igdisadiyyat ve Hayat," ISSN 0207-3021, Baku, **2007,** *9,* 42–48.
33. Aliyev Azad, G. *Construction of economic-mathematical model with regard to unaccounted factors influence function.* Proceedings of Republic scientific-practical conference of postgraduate students and young scientists. Baku, **1998,** *2,* 243–244.
34. Aliyev Azad, G. *Principle of economic process certainty in finite-dimensional vector space.* "Igdisadiyyat ve Hayat," ISSN 0207-302*1,* Baku, **2007,** *8,* 43–47.
35. Aliyev Azad, G. *Economic-mathematical methods and models with regard to incomplete information.* Ed. "Elm," ISBN 5-8066-1487-5, Baku, **2002,** 288.
36. Aliyev Azad, G.; *Theoretical bases of economic-mathematical simulation at uncertainty conditions,* National Academy of Sciences of Azerbaijan, Baku-2011, "Information Technologies," ISBN 995221056-9, **2011,** 338.
37. Aliyev Azad, G. *Economical and mathematical models subject to complete information.* Ed. Lap Lambert Akademic Publishing. ISBN-978-3-659-30998-4, Berlin, **2013,** 316.

CHAPTER 3

PIECEWISE LINEAR ECONOMIC-MATHEMATICAL MODELS WITH REGARD TO UNACCOUNTED FACTORS INFLUENCE IN 3-DIMENSIONAL VECTOR SPACE

CONTENTS

In Chapter 2, theory of construction of piecewise-linear economic mathematical models with regard to unaccounted factors influence in finite-dimensional vector space was developed. A method for predicting economic process and controlling it at uncertainty conditions, and a way for defining the economic process control function in m-dimensional vector space, were suggested.

In this chapter, we give a number of practically important piecewise-linear economic-mathematical models with regard to unaccounted parameters influence factor in their-dimensional vector space. And by means of 2- and 3-component piecewise-linear models suggest an appropriate method of multivariant prediction of economic process in subsequent stages and its control then at uncertainty conditions in 3-dimensional vector space.

3.1 2-COMPONENT PIECEWISE-LINEAR ECONOMIC-MATHEMATICAL MODEL AND METHOD OF MULTIVARIATE PREDICTION OF ECONOMIC PROCESS WITH REGARD TO UNACCOUNTED FACTORS INFLUENCE IN 3-DIMENSIONAL VECTOR SPACE

Given a statistical table describing some economic process in the form of the points (vector) set $\{\vec{a}_n\}$ of 3-dimensional vector space R_3. Here the numbers a_{ni} are the coordinates of the vector $\vec{a}_n$ ($a_{n1}, a_{n2}, a_{n3}, \ldots\ldots a_{ni}$). With the help of the points (vectors) $\vec{a}_n$ represent the set of statistical points in the vector form in the form of 2-component piecewise-linear function [1–6]:

$$\vec{z}_1 = \vec{a}_1 + \mu_1(\vec{a}_2 - \vec{a}_1) \tag{1}$$

$$\vec{z}_2 = \vec{a}_2 + \mu_2(\vec{a}_3 - \vec{a}_2) \tag{2}$$

Here $\vec{z}_1 = \vec{z}_1(z_{11}, z_{12}, z_{13})$ and $\vec{z}_2 = \vec{z}_2(z_{21}, z_{22}, z_{23})$ are the equations of the first and second piecewise-linear straight lines on 3-dimensional vector space; the vectors $\vec{a}_1(a_{11}, a_{12}, a_{13})$, $\vec{a}_2 = \vec{a}_2(a_{21}, a_{22}, a_{23})$ and $\vec{a}_3 = \vec{a}_3(a_{31}, a_{32}, a_{33})$ are the given points (vectors) in 3-dimensional space; $\mu_1 \geqslant 0$ and $\mu_1 \geqslant 0$ are arbitrary parameters of the first and second piecewise-linear straight lines. And it holds the equality $\lambda_1 + \mu_1 = 1$ and $\lambda_2 + \mu_2 = 1$; $\alpha_{1,2}$ is the angle between the piecewise-linear straight lines; k_1 is the intersection point between the first and second straight lines (Fig. 1).

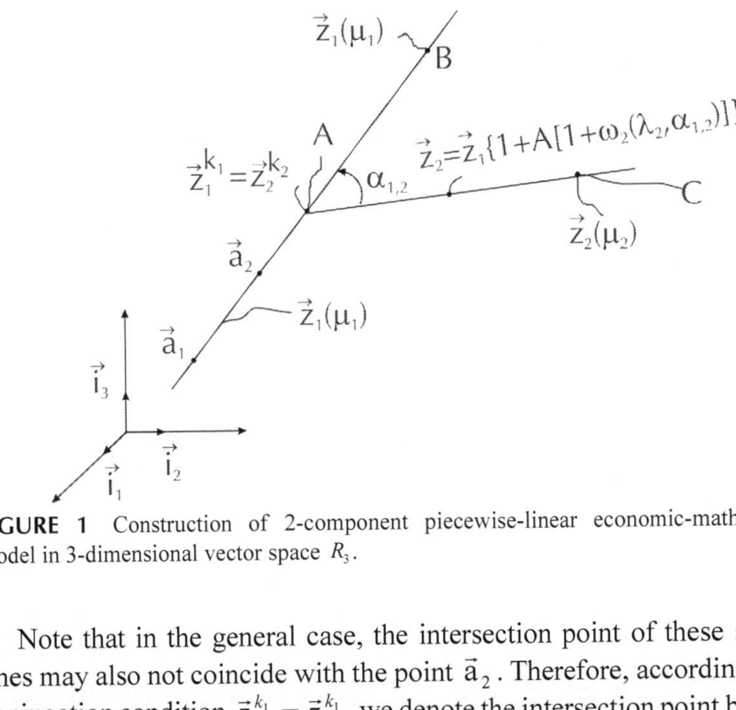

FIGURE 1 Construction of 2-component piecewise-linear economic-mathematical model in 3-dimensional vector space R_3.

Note that in the general case, the intersection point of these straight lines may also not coincide with the point $\vec{a}_2$. Therefore, according to the conjugation condition $\vec{z}_1^{k_1} = \vec{z}_2^{k_1}$, we denote the intersection point between the first and second piecewise-linear straight lines in 3-dimensional vector space by $\vec{z}_1^{k_1}$ (Fig. 1). Allowing for this fact, we write the equation of the second piecewise-linear straight line in the form Eq. (10):

$$\vec{z}_2 = \vec{z}_1^{k_1} + \mu_2(\vec{a}_3 - \vec{z}_1^{k_1}) \tag{3}$$

where the value $\vec{z}_1^{k_1}$ is the value of the point (vector) of the first piecewise-linear straight line at the k_1-th intersection point and equals:

$$\vec{z}_1^{k_1} = \vec{a}_1 + \mu_1^{k_1}(\vec{a}_2 - \vec{a}_1) \tag{4}$$

In particular case, $\vec{z}_1^{k_1} = \vec{a}_2$ for $\mu_1^{k_1} = 1$. In this case, the intersection point $\vec{z}_1^{k_1}$ coincides with the point $\vec{a}_2$.

Now, according to Eqs. (70)–(76) of Chapter 2, the vector equation for the points of the second piecewise-linear straight line depending on the vector function of the first piecewise-linear straight line $\vec{z}_1$ and introduced unaccounted parameters influence spatial function $\omega_2(\lambda_2, \alpha_{1,2})$ in 3-dimensional vector space is written in the form (Fig. 1) [7–9]:

$$\vec{z}_2 = \vec{z}_1\{1 + A\,[1 + \omega_2(\lambda_2, \alpha_{1,2})]\} \tag{5}$$

Here

$$A = (\mu_1^{k_1} - \mu_1)\frac{|\vec{a}_2 - \vec{a}_1||\vec{z}_1^{k_1} - \vec{a}_1|}{\vec{z}_1(\vec{z}_1^{k_1} - \vec{a}_1)} \tag{6}$$

$$\lambda_2 = \frac{\mu_2}{\mu_1^{k_1} - \mu_1} \cdot \frac{|\vec{z}_1 - \vec{z}_1^{k_1}||\vec{a}_3 - \vec{z}_1^{k_1}|}{\vec{z}_1(\vec{z}_1 - \vec{z}_1^{k_1})} \cdot \frac{\vec{z}_1(\vec{z}_1^{k_1} - \vec{a}_1)}{|\vec{a}_2 - \vec{a}_1||\vec{z}_1^{k_1} - \vec{a}_1|} \tag{7}$$

$$\omega_2(\lambda_2, \alpha_{1,2}) = \lambda_2 cos\alpha_{1,2} \tag{8}$$

$$\mu_2 = (\mu_1 - \mu_1^{k_1})\frac{(\vec{a}_3 - \vec{z}_1^{k_1})(\vec{a}_2 - \vec{a}_1)}{(\vec{a}_3 - \vec{z}_1^{k_1})^2}, \text{ for } \mu_1 > \mu_1^{k_1} \tag{9}$$

Eq. (9) is the mathematical relation between arbitrary parameters μ_2 and μ_1. For the second piecewise-linear straight line, representing a straight line restricted with one end, condition Eq. (9) will hold for all $\mu_1 > \mu_1^{k_1}$. Furthermore, for the second intersection point k_2, i.e., for $\mu_2 = \mu_2^{k_2}$, the appropriate value of the parameter μ_1 will be determined as follows:

$$\mu_1^{k_2} = \mu_1^{k_1} + \frac{(\vec{a}_3 - \vec{z}_1^{k_1})^2}{(\vec{a}_3 - \vec{z}_1^{k_1})(\vec{a}_2 - \vec{a}_1)} \mu_2^{k_2} \tag{10}$$

The value of $cos\ \alpha_{1,2}$ between the first and second piecewise-linear straight lines is determined by means of the scalar product of 2 vectors of the form (Fig. 1):

$$cos\alpha_{1,2} = \frac{\vec{AB} \cdot \vec{AC}}{|\vec{AB}| \cdot |\vec{AC}|} = \frac{(\vec{z}_1 - \vec{z}_1^{k_1})(\vec{z}_2 - \vec{z}_1^{k_1})}{|\vec{z}_1 - \vec{z}_1^{k_1}| \cdot |\vec{z}_2 - \vec{z}_1^{k_1}|} \tag{11}$$

By calculating the values of $cos\ \alpha_{1,2}$ we can use any values of arbitrary parameters μ_1 and μ_2.

Thus, in 3-dimensional vector space, determining the points (vectors):

$$\vec{a}_1 = \sum_{m=1}^{3} a_{1m}\vec{i}_m, \quad \vec{a}_2 = \sum_{m=1}^{3} a_{2m}\vec{i}_m, \quad \vec{a}_3 = \sum_{m=1}^{3} a_{3m}\vec{i}_m$$

$$\vec{z}_1^{k_1} = \sum_{m=1}^{3} z_{1m}^{k_1}\vec{i}_m = \sum_{m=1}^{3} [a_{1m} + \mu_1^{k_1}(a_{2m} - a_{1m})]\vec{i}_m \tag{12}$$

Eq. (5) will represent an equation for the second vector straight line $\vec{z}_2 = \vec{z}_2(\mu_1, \omega_2)$ depending on the unaccounted parameter influence function $\omega_2(\lambda_2, \alpha_{1,2})$ and arbitrary parameter $\mu_1 \geqslant \mu_1^{k_1}$.

Represent the vector equation for the second piecewise-linear straight line Eq. (5) in the coordinate form. For that take into account that in 3-dimensional space the vectors of the first and second piecewise-linear straight lines in the coordinate form are represented in the form:

$$\vec{z}_1 = \sum_{m=1}^{3} z_{1m}\vec{i}_m \quad \text{and} \quad \vec{z}_2 = \sum_{m=1}^{3} z_{2m}\vec{i}_m \tag{13}$$

In this case, the coordinates of the vector $\vec{z}_2$ Eq. (5), i.e., z_{2m} will be expressed by the coordinates of the first piecewise-linear vector z_{1m}, spatial vector λ_2 and the unaccounted parameter influence function $\omega_2(\lambda_2, \alpha_{1,2})$, in the form:

$$z_{2m} = \{1 + A[1 + \omega_2(\lambda_2, \alpha_{1,2})]\} z_{1m}, \text{ for } m = 1,2,3 \tag{14}$$

Here the coordinate notation of the coefficients A, λ_2 and $\omega_2(\lambda_2, \alpha_{1,2})$, by Eqs. (6)–(9), will be of the form:

$$\lambda_2 = \frac{\mu_2}{\mu_1^{k_1} - \mu_1} \cdot \frac{\sqrt{\sum_{i=1}^{3} \{a_{3i} - [a_{1i} + \mu_1^k (a_{2i} - a_{1i})]\}^2}}{\sqrt{\sum_{i=1}^{3} (a_{2i} - a_{1i})^2}} \tag{15}$$

$$\lambda_2 = \frac{\mu_2}{\mu_1^{k_1} - \mu_1} \cdot \frac{\sqrt{\sum_{i=1}^{3} \{a_{3i} - [a_{1i} + \mu_1^k (a_{2i} - a_{1i})]\}^2}}{\sqrt{\sum_{i=1}^{3} (a_{2i} - a_{1i})^2}} \tag{16}$$

$$\omega_2(\lambda_2, \alpha_{1,2}) = \lambda_2 cos\alpha_{1,2} \tag{17}$$

$$\mu_2 = (\mu_1 - \mu_1^{k_1}) \frac{\sum_{i=1}^{3} (a_{2i} - a_{1i})[a_{3i} - a_{1i} - \mu_1^{k_1} (a_{2i} - a_{1i})]}{\sum_{i=1}^{3} [a_{3i} - a_{1i} - \mu_1^{k_1} (a_{2i} - a_{1i})]^2},$$

for

$$\mu_1 \geq \mu_1^{k_1} \tag{18}$$

$$\mu_2^{k_2} = \frac{\sum_{i=1}^{3}(a_{2i} - a_{1i})[a_{3i} - a_{1i} - \mu_1^{k_1}(a_{2i} - a_{1i})]}{\sum_{i=1}^{3}[a_{3i} - a_{1i} - \mu_1^{k_1}(a_{2i} - a_{1i})]^2}(\mu_1^{k_2} - \mu_1^{k_1}) \tag{19}$$

Now, for the case economic process represented in the form of 2-component piecewise-linear economic-mathematical model, investigate the prediction and control of such a process on the subsequent $y_n = y_n(t, \lambda_n)$ small volume of 3-dimensional vector space with regard to unaccounted parameter influence function $\omega_2(\lambda_2, \alpha_{1,2})$. And the value of the unaccounted parameter $\omega_2(\lambda_2, \alpha_{1,2})$ function is assumed to be known [10, 11].

A method for constructing a predicting vector function of economic process $\vec{Z}_{N+1}(\beta)$ with regard to the introduced unaccounted parameters influence predicting function $\Omega_{N+1}(\lambda_{N+1}, \alpha_{N,N+1})$ in m-dimensional vector space, represented by Eqs. (19)–(25) was developed in Section 2.3, Chapter 2. Apply this method to the case of the given 2-component piecewise-linear economic model 3-dimensional vector space. It will be of the form:

$$\vec{Z}_3(1) = \vec{z}_1\{1 + A[1 + \omega_2(\lambda_2^{k_2}, \alpha_{1,2}) + \Omega_3(\lambda_3, \alpha_{2,3})]\} \tag{20}$$

Where

$$\omega_2(\lambda_2^{k_2}, \alpha_{1,2}) = \lambda_2^{k_2} \cdot cos\alpha_{1,2} =$$

$$= \frac{\mu_2^{k_2}}{\mu_1^{K_1} - \mu_1} \cdot \frac{|\vec{z}_1 - \vec{z}_1^{k_1}| \cdot |\vec{a}_3 - \vec{z}_1^{k_1}|}{\vec{z}_1(\vec{z}_1 - \vec{z}_1^{k_1})} \cdot \frac{\vec{z}_1(\vec{z}_1^{k_1} - \vec{a}_1)}{|\vec{a}_2 - \vec{a}_1| \cdot |\vec{z}_1^{k_1} - \vec{a}_1|} cos\alpha_{1,2} \tag{21}$$

$$\mu_2 = (\mu_1 - \mu_1^{k_1}) \cdot \frac{(\vec{a}_2 - \vec{a}_1)(\vec{a}_3 - \vec{z}_1^{k_1})}{(\vec{a}_3 - \vec{z}_1^{k_1})^2}, \quad \mu_1 \geq \mu_1^{k_1} \tag{22}$$

$$A = (\mu_1^{k_1} - \mu_1) \cdot \frac{|\vec{a}_2 - \vec{a}_1| \cdot |\vec{z}_1^{k_1} - \vec{a}_1|}{\vec{z}_1(\vec{z}_1^{k_1} - \vec{a}_1)} \tag{23}$$

$$\Omega_3(\lambda_3, \alpha_{2,3}) = \lambda_3 \cdot cos\alpha_{2,3} \tag{24}$$

$$\lambda_3 = \frac{\mu_3}{\mu_1^{k_1} - \mu_1} \cdot \frac{\left|\vec{z}_2 - \vec{z}_2^{k_2}\right| \cdot \left|\vec{a}_4(1) - \vec{z}_2^{k_2}\right|}{\vec{z}_1(\vec{z}_2 - \vec{z}_2^{k_2})} \cdot \frac{\vec{z}_1(\vec{z}_1^{k_1} - \vec{a}_1)}{\left|\vec{a}_2 - \vec{a}_1\right| \cdot \left|\vec{z}_1^{k_1} - \vec{a}_1\right|} \tag{25}$$

$$\mu_3 = (\mu_2 - \mu_2^{k_2}) \cdot \frac{(\vec{a}_3 - \vec{z}_1^{k_1})(\vec{a}_4(1) - \vec{z}_2^{k_2})}{(\vec{a}_4(1) - \vec{z}_2^{k_2})^2}, \tag{26}$$

$$\mu_2 \geq \mu_2^{k_2},\ \mu_3 \geq 0$$

Here, according to Eq. (17), the vector $\vec{a}_4(\beta)$ is of the form:

$$\vec{a}_4(1) = a_{41}(1)\vec{i}_1 + a_{42}(1)\vec{i}_2 + a_{43}(1)\vec{i}_3 = \sum_{m=1}^{3} a_{4m}(1) \cdot \vec{i}_m \tag{27}$$

And the coordinates of a_{42} and a_{43} are expressed by the arbitrarily given coordinate $a_{41} > z_{21}^{k_2}$ in the form:

$$C = \frac{a_{41}(1) - z_{21}^{k_2}}{a_{21} - a_{11}} = \frac{a_{42}(1) - z_{22}^{k_2}}{a_{22} - a_{12}} = \frac{a_{43}(1) - z_{23}^{k_2}}{a_{23} - a_{13}} \tag{28}$$

Hence:

$$a_{42}(1) = z_{22}^{k_2} + \frac{a_{22} - a_{12}}{a_{21} - a_{11}}(a_{41}(1) - z_{21}^{k_2})$$

$$a_{43}(1) = z_{23}^{k_3} + \frac{a_{23} - a_{13}}{a_{21} - a_{11}}(a_{41}(1) - z_{21}^{k_2}) \tag{29}$$

Here the coefficients a_{2m}, a_{1m} and $z_{2m}^{k_2}$ are the coordinates of the vectors $\vec{a}_1$, $\vec{a}_2$, $\vec{z}_2^{k_2}$ in 3-dimensional vector space and equal:

$$\vec{a}_1 = \sum_{m=1}^{3} a_{1m} \cdot \vec{i}_m \, , \; \vec{a}_2 = \sum_{m=1}^{3} a_{2m} \cdot \vec{i}_m \, , \; \vec{z}_2^{k_2} = \sum_{m=1}^{3} \vec{z}_{2m}^{k_2} \vec{i}_m \tag{30}$$

Note that in the vectors $\vec{Z}_3(1)$ and $\vec{a}_4(1)$ the index (1) in the brackets means that the vector $\vec{Z}_3(1)$ is parallel to the first piecewise-linear vector function $\vec{z}_1$. This means that the economic process beginning with the point $\vec{z}_2^{k_2}$ will hold by the scenario of the first piecewise-linear equation (Fig. 2).

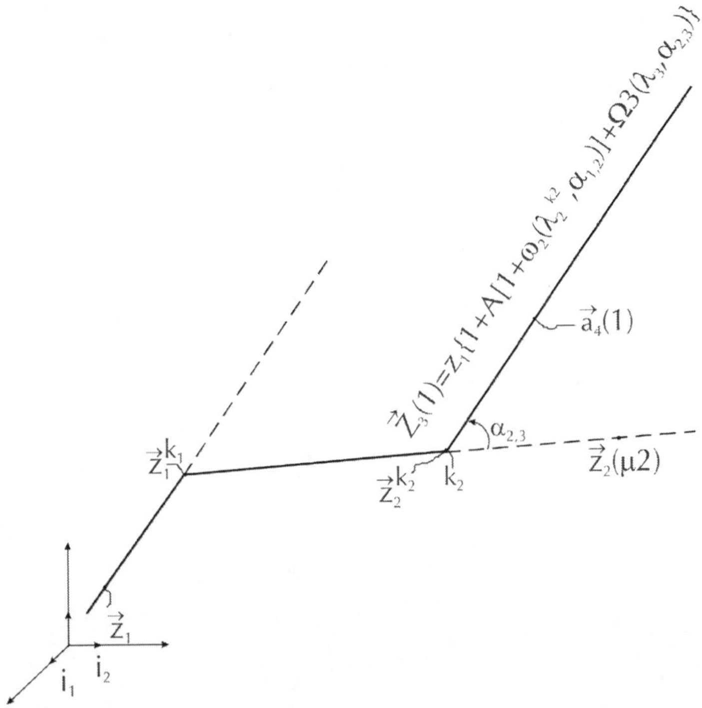

FIGURE 2 Construction of the predicting vector function $\vec{Z}_3(\beta)$ with regard to unaccounted parameter influence predicting function $\Omega_3(\lambda_3, \alpha_{2,3})$ on the base of 2-component economic-mathematical model in 3-dimensional vector space R_3.

The expression of $cos\alpha_{2.3}$ corresponding to the cosine of the angle between the second piecewise-linear straight line $\vec{z}_2$ and the predicting third vector straight line $\vec{z}_3(1)$ on the base of the scalar product of 2 vectors, is represented in the form (Fig. 2):

$$cos\alpha_{2,3} = \frac{(\vec{z}_2 - \vec{z}_2^{k_2}) \cdot (\vec{a}_4(1) - \vec{z}_2^{k_2})}{\left|\vec{z}_2 - \vec{z}_2^{k_2}\right| \left|\vec{a}_4(1) - \vec{z}_2^{k_2}\right|} \tag{31}$$

3.1.1 METHOD OF NUMERICAL CALCULATION OF 2-COMPONENT ECONOMIC-MATHEMATICAL MODEL AND DEFINITION OF PREDICTING VECTOR FUNCTION WITH REGARD TO UNACCOUNTED FACTORS INFLUENCE IN 3-DIMENSIONAL VECTOR SPACE

Below we have given the numerical construction of a 2-component piecewise-linear economic mathematical model, and by means of the given model will determine the predicting function on the subsequent third small volume of the investigated economic process in 3-dimensional vector space [11, 12].

Given a statistical table describing some economic process in the form of the points (vectors) set $\{\vec{a}_n\}$ in 3-dimensional vector space R_3. Represent the set of vectors $\{\vec{a}_n\}$ of statistical values in the form of adjacent 2-component piecewise-linear vector equation of the form Eq. (1):

$$\vec{z}_1 = \vec{a}_1 + \mu_1(\vec{a}_2 - \vec{a}_1) \tag{32}$$

$$\vec{z}_2 = \vec{z}_1^{k_1} + \mu_2(\vec{a}_3 - \vec{z}_1^{k_1}) \tag{33}$$

where $\vec{z}_1 = \vec{z}_1(z_{11}, z_{12}, z_{13})$ and $\vec{z}_2 = \vec{z}_2(z_{21}, z_{22}, z_{23})$ are the equations of the first and second piecewise-linear straight lines in 3-dimensional vector space; the vectors $\vec{a}_1(a_{11}, a_{12}, a_{13})$, $\vec{a}_2 = \vec{a}_2(a_{21}, a_{22}, a_{23})$ and $\vec{a}_3 = \vec{a}_3(a_{31}, a_{32}, a_{33})$ are given points (vectors) in 3-dimensional space of the form:

$$\vec{a}_1 = \vec{i}_1 + \vec{i}_2 + \vec{i}_3 \quad \vec{a}_2 = 3\vec{i}_1 + 2\vec{i}_2 + 4{,}5\vec{i}_3$$

$$\vec{a}_3 = 6\vec{i}_1 + 4\vec{i}_2 + 7\vec{i}_3 \tag{34}$$

$\mu_1 \geqslant 0$ and $\mu_2 \geqslant 0$ are arbitrary parameter. Substituting Eq. (34) in Eq. (32), the coordinate form of the vector equation of the first vector straight line will accept the form:

$$\vec{z}_1 = (1+2\mu_1)\vec{i}_1 + (1+\mu_1)\vec{i}_2 + (1+3{,}5\mu_1)\vec{i}_3 \tag{35}$$

As the intersection point of 2 straight lines $\vec{z}_1^{k_1}$ that should satisfy the conjugation condition $\vec{z}_1^{k_1} = \vec{z}_2^{k_1}$ may also not coincide with the point $\vec{a}_2$, then its appropriate value of the parameter μ_1 will be $\mu_1^{k_1} \geq 1$. In this connection, in numerical calculation, we accept the value of the parameter $\mu_1^{k_1}$ for the intersection point between piecewise-linear straight lines equal 1.5, i.e., $\mu_1^{k_1} = 1{,}5$. Then the value of the intersection point $\vec{z}_1^{k_1}$ Eq. (35) will equal:

$$\vec{z}_1^{k_1} = 4\vec{i}_1 + 2{,}5\vec{i}_2 + 6{,}25\vec{i}_3 \tag{36}$$

By Eq. (5) the equation of the second straight line in the vector form is expressed by the vector equation of the first piecewise-linear straight line $\vec{z}_1$ of the form Eq. (35) and the unaccounted parameter function $\omega_2(\lambda_2, \alpha_{1,2})$ in the form:

$$\vec{z}_2 = \vec{z}_1\{1 + A[1 + \omega_2(\lambda_2, \alpha_{1,2})]\} \tag{37}$$

Here the coefficient A and the unaccounted parameter function $\omega_2(\lambda_2, \alpha_{1,2})$ of the economic process will be of the form Eqs. (6)–(9) and (11):

$$A = (\mu_1^{k_1} - \mu_1)\frac{|\vec{a}_2 - \vec{a}_1||\vec{z}_1^{k_1} - \vec{a}_1|}{\vec{z}_1(\vec{z}_1^{k_1} - \vec{a}_1)} \quad \text{for } \mu_1 \geqslant \mu_1^{k_1} = 1{,}5 \tag{38}$$

$$\omega_2(\lambda_2^{k_2}, \alpha_{1,2}) = \lambda_2^{k_2} \cos\alpha_{1,2} \quad (39)$$

$$\lambda_2 = \frac{\mu_2^{k_2}}{\mu_1^{k_1} - \mu_1} \frac{|\vec{z}_1 - \vec{z}_1^{k_1}||\vec{a}_3 - \vec{z}_1^{k_1}|}{\vec{z}_1(\vec{z}_1 - \vec{z}_1^{k_1})} \frac{\vec{z}_1(\vec{z}_1^{k_1} - \vec{a}_1)}{|\vec{a}_2 - \vec{a}_1||\vec{z}_1^{k_1} - \vec{a}_1|} \quad (40)$$

$$cos\alpha_{1,2} = \frac{(\vec{z}_1 - \vec{z}_1^{k_1})(\vec{z}_2 - \vec{z}_1^{k_1})}{|\vec{z}_1 - \vec{z}_1^{k_1}||\vec{z}_2 - \vec{z}_1^{k_1}|} \quad (41)$$

Here the parameter μ_2 corresponding to the points of the second piece-wise-linear straight line is connected with the appropriate parameter μ_1 by Eq. (9). Here for the values $\mu_1 \geq \mu_1^{k_1} = 1{,}5$.

In Eq. (41) the vector $\vec{z}_2$ is calculated by Eq. (33) for any value of μ_2 in the interval $0 \leq \mu_2 \leq 1$, and the vector $\vec{z}_1$ is of the form Eq. (32) for any value of $\mu_2 \geq \mu_1^{k_1}$. By calculating the value of the expression $Cos\alpha_{1,2}$ by Eq. (41), the value of $\vec{z}_1$ may be calculated for the value of $\vec{a}_3$ or for μ_2 that corresponds to the value of the second intersection point k_2, i.e., for $\mu \geq \mu_2^{k_2}$.

Substituting the value of the parameter $\mu_1^{k_1} = 1{,}5$, and also Eq. (34) in Eq. (9), set up a numerical relation between the parameters μ_2 and μ_1 in the form:

$$\mu_2 = 1{,}1927(\mu_1 - 1{,}5) \text{ for } \mu_1 \geq 1{,}5;\ 0 \leq \mu_2 \leq \mu_2^{k_2} > 1 \quad (42)$$

Thus, Eq. (42) is the numerical representation of mathematical relation between the parameters μ_1 and μ_2. Defining any value of $\mu_2 \geq 0$ by Eq. (42), it is easy to determine the appropriate value of the parameter μ_1. From Eq. (42) it will follow:

$$\mu_1 = 1{,}5 + 0{,}8384\mu_2 \quad (42a)$$

Calculate the values of the coefficient A, the unaccounted parameter function $\omega_2(\lambda_2, \alpha_{1,2})$ of economic process and $Cos\alpha_{1,2}$. For that, substituting Eqs. (34)–(36) in Eq. (42), and also the numerical value of the parameter $\mu_1^{k_1} = 1{,}5$ in Eqs. (38)–(41), define the numerical values of A, $\omega_2(\lambda_2, \alpha_{1,2})$ and $Cos\alpha_{1,2}$ for $\mu_1 \geq 1{,}5$ in the form:

$$A = -(\mu_1 - 1{,}5)\frac{25{,}8751}{9{,}75 + 25{,}875 \cdot \mu_1} \tag{43}$$

$$\lambda_2 = 0{,}1208\frac{9{,}75+25{,}875\mu_1}{9{,}75+19{,}375\mu_1 0-17{,}25\mu_1^2}\sqrt{38{,}8125-51{,}25\mu_1+17{,}25\mu_1^2} \tag{44}$$

$$cos\alpha_{1,2} = 0.8495 \tag{45}$$

Numerical values of A and λ_2 for the second intersection point, i.e., for $\mu_1 = 3{,}1768$ calculated by Eqs. (43) and (44) will be equal to:

$$A(3{,}1768) = -0{,}4719, \quad \lambda_2 = (3{,}1768) = -0{,}7495$$

Substituting Eqs. (35), (43)–(45) in Eq. (37), find the equation of the second vector straight line in the vector form depending on the vector function of the first piecewise-linear straight line and appropriate for the second linear straight line of the parameter $\mu_1 \geq 1{,}5$ in the form (Fig. 3):

$$\vec{z}_2 = \varphi_0(\mu_1)\cdot\vec{z}_1 = \varphi_0(\mu_1)\cdot[(1+2\mu_1)\vec{i}_1 + (1+\mu_1)\vec{i}_2 + (1+3{,}5\mu_1)\vec{i}_3]$$

for

$$\mu_1 \geq 1{,}5 \tag{46}$$

Here

$$\varphi_0(\mu_1) = 1-(\mu_1 - 1{,}5)\cdot\frac{25{,}8751}{9{,}75+25{,}875\cdot\mu_1}\left[1+ \right.$$

$$\left. +0{,}1026\frac{9{,}75+25{,}875\cdot\mu_1}{9{,}75+19{,}375\cdot\mu_1-17{,}25\mu_1^2}\sqrt{38{,}8125-51{,}75\cdot\mu_1 + +17{,}25\mu_1^2}\right] \tag{47}$$

Numerical values $\varphi_0(\mu_1)$ at the second intersection point, i.e., for $\mu_1 = 3{,}1768$ will equal:

$$\varphi_0(3{,}1768) = 0{,}8297$$

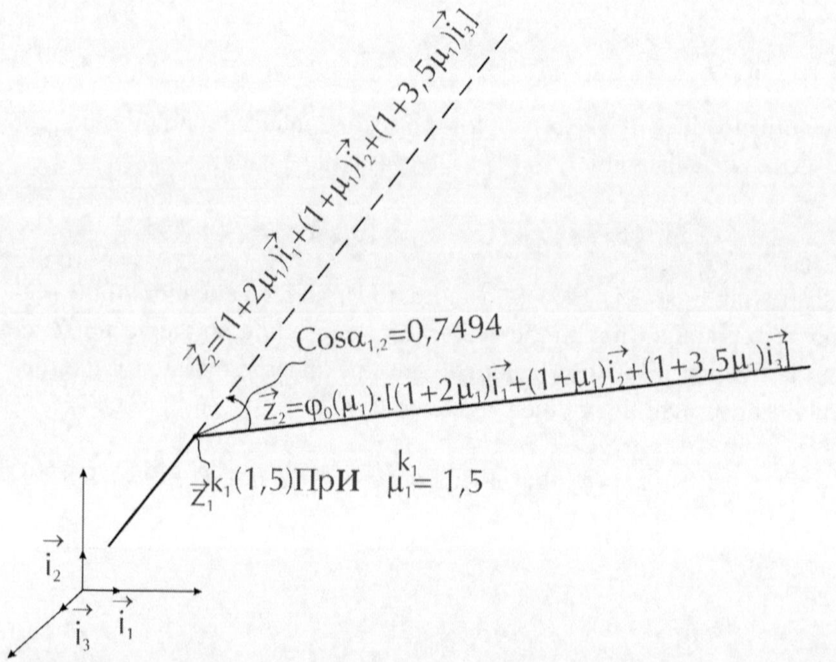

FIGURE 3 Numerical representation of 2-component piecewise-linear economic-mathematical model in 3-dimensional vector space R_3.

Now investigate the problem of prediction and control of economic process in the subsequent $\Delta V_3(x_1, x_2, x_3)$ volume of 3-dimensiona vector space with regard to unaccounted parameters factor that hold on preceding states of the process [11, 13].

Above for the case of 2-component piecewise-linear straight line it was numerically constructed the second vector straight line (46) depending on an arbitrary parameter μ_1 and unaccounted parameter influence space function $\omega_2(\lambda_2, \alpha_{1,2})$. On the other hand, for the 2-component case economic process a predicting vector function $\vec{Z}_3(1)$ with regard to the introduced unaccounted parameter influence predicting function $\Omega_3(\lambda_3, \alpha_{2,3})$ was suggested in the form:

$$\vec{Z}_3(1) = \vec{z}_1\{1 + A[1 + \omega_2(\lambda_2^{k_2}, \alpha_{1,2}) + \Omega_3(\lambda_3, \alpha_{2,3})]\} \tag{48}$$

Here the coefficient A, the unaccounted parameter function $\omega_2(\lambda_2^{k_2}, \alpha_{1,2})$, , and also the unaccounted parameter predicting function $\Omega_3(\lambda_3, \alpha_{2,3})$ are of the form Eqs. (21)–(26) define numerical values of these expressions.

As the economic process predicting function $\vec{Z}_3(1)$ is the third piece-wise-linear function, at first we define the value of the vector function $\vec{z}_2^{k_2}$ at the second intersection point k_2. The parameter μ_2 acting on the segment of the second piecewise-linear straight line changes in the interval $0 \le \mu_2 \le \mu_2^{k_2} \ge 1$. Here the value of the parameter $\mu_2^{k_2}$ belongs to the intersection point between the second and third straight lines. According to approximation of statistical points, this point should be defined. Therefore, giving the value of the parameter $\mu_2^{k_2}$ at the second intersection point k_2, define from Eq. (9) the appropriate value of the parameter $\mu_1^{k_2}$, in the form:

$$\mu_1^{k_2} = \mu_1^{k_1} + \mu_2^{k_2} \frac{(\vec{a}_3 - \vec{z}_1^{k_1})^2}{(\vec{a}_3 - \vec{z}_1^{k_1})(\vec{a}_2 - \vec{a}_1)} \tag{49}$$

For conducting numerical calculation we accept $\mu_2^{k_2} = 2$. For the value of the parameter $\mu_2^{k_2} = 2$, we define the appropriate numerical value of the parameter μ_1, that will be denoted by $\mu_1^{k_2}$, from Eq. (49) or Eq. (42). It will equal:

$$\mu_1^{k_2} = 3{,}1768 \tag{50}$$

Thus, we established the range of the parameter μ_1 corresponding to the change of the parameter μ_2 of the segment of the second piecewise-linear straight line, in the form:

$$1{,}5 \le \mu_1 \le 3{,}1768 \text{ for } 0 \le \mu_2 \le \mu_2^{k_2} = 2 \tag{51}$$

Though Eq. (49) is valid for the values of the parameter $\mu_2 \ge 2$ as well.

In this case, the value of the prediction function $\vec{Z}_3^{k_2}(1)$ at the intersection point k_2 i.e., for $\mu_3 = 0$, $\mu_2 = 2$, $\mu_1^{k_2} = 3{,}1768$ coincides with the value of the function of the second piecewise-linear straight line:

$$\vec{Z}_3^{k_2}(1) = \vec{z}_2^{k_2} \tag{52}$$

Note that at the intersection point k_2, i.e., for $\mu_2^{k_2} = 2$, $\mu_3^{k_2} = 0$ the unaccounted parameters influence predicting function $\Omega_3(\lambda_3, \alpha_3) = 0$.

But the function $\vec{z}_2$ has the form (46). Therefore, it suffices to substitute to Eq. (46) the value of the parameter $\mu_1^{k_2} = 3{,}1768$ that will be defined both as the value of the predicting function $\vec{Z}_3^{k_2}(1)$ at the initial point $\mu_2 = 2$ $\mu_3 = 0$ of the third vector straight line and the value of the point $\vec{z}_2^{k_2}$ at the final point of the second piecewise-linear straight line at the point κ_2, in the form:

$$Z_3^{k_2}(1)\Big|_{\mu_1 = 3{,}1768} = 6{,}1013\vec{i}_1 + 3{,}4655\vec{i}_2 + 10{,}055\vec{i}_3$$

for

$$\mu_2 = 2,\ \mu_1^{k_2} = 3{,}1768,\ \mu_3 = 0 \tag{53}$$

Calculate the point $\vec{a}_4(1)$. For that give in an arbitrary form 1 of the coordinates of the vector $\vec{a}_4(1)$, for instance, the coordinate $a_{41}(1)$, and by Eq. (29) calculate the remaining coordinates of the vector $\vec{a}_4(1)$. Furthermore, $a_{41}(1)$ is given so that $a_{41}(1)$ were greater than the coordinates $z_{21}^{k_2} = 5{,}8411$. Therefore accept the value $a_{41}(1)$=6,5. In this case, substituting Eqs. (34) and (53) in Eq. (29), define the vector $\vec{a}_4(1)$ in the coordinate form depending on an arbitrarily given value of $a_{41}(1)$ in the form:

$$\vec{a}_4(1) = a_{41}\vec{i}_1 + (1{,}3707 + \frac{1}{3}a_{41})\vec{i}_2 + (-3{,}5163 + 2{,}25a_{41})\vec{i}_3 \tag{54}$$

For the value $a_{41}(1)$=six,5, the vector accepts the form $\vec{a}_4(1)$:

$$\vec{a}_4(1) = 6{,}5\vec{i}_1 + 3{,}5374\vec{i}_2 + 11{,}1087\vec{i}_3 \tag{55}$$

For numerical definition of the coefficient A the unaccounted parameter function $\omega_2(\lambda_2^{k_2}, \alpha_{1,2})$ and also the unaccounted parameter predicting function $\Omega_3(\lambda_3, \alpha_{2,3})$ allowing for Eqs. (34)–(36), (42), (47), (9) and (53) conduct the following calculations:

$$1)\ \left|\vec{z}_1^{k_1} - \vec{a}_1\right| = \left|4\vec{i}_1 + 2{,}5\vec{i}_2 + 6{,}25\vec{i}_3 - \vec{i}_1 - \vec{i}_2 - \vec{i}_3\right| = 6{,}23 \tag{56}$$

$$2)\ \left|\vec{a}_2 - \vec{a}_1\right| = \left|2\vec{i}_1 + \vec{i}_2 + 3{,}5\vec{i}_3\right| = 4{,}1533 \tag{57}$$

$$3)\ \vec{z}_1(\vec{z}_1^{k_1} - \vec{a}_1) = \left[(1+2\mu_1)\vec{i}_1 + (1+\mu_1)\vec{i}_2 + (1+3{,}5\mu_1)\vec{i}_3\right](3\vec{i}_1 +$$

$$+1{,}5\vec{i}_2 + 5{,}25\vec{i}_3) = 9.75 + 25{,}875\mu_1 = A_1(\mu_1) \tag{58}$$

$$4)\ \left(\vec{z}_2 - \vec{z}_2^{k_2}\right) = \{\varphi_0(\mu_1)\left[(1+2\mu_1)\vec{i}_1 + (1+\mu_1)\vec{i}_2 + (1+3{,}5\mu_1)\vec{i}_3\right] -$$

$$-5{,}8411\vec{i}_1 - 3{,}3177\vec{i}_2 - 9{,}6262\vec{i}_3\} =$$

$$= \left[\varphi_0(1+2\mu_1) - 5{,}8411\right]\vec{i}_1 + \left[\varphi_0(1+\mu_1) - 3{,}3177\right]\vec{i}_2 +$$

$$+\left[\varphi_0(1+3{,}5\mu_1) - 9{,}6262\right]\vec{i}_3 \tag{59}$$

$$5)\ \left|\vec{z}_2 - \vec{z}_2^{k_2}\right| = \left|\ \varphi_0(\mu_1)\left[(1+2\mu_1)\vec{i}_1 + (1+\mu_1)\vec{i}_2 + (1+3{,}5\mu_1)\vec{i}_3\right] -\right.$$

$$\left. -5{,}8411\vec{i}_1 - 3{,}3177\vec{i}_2 - 9{,}6262\vec{i}_3\ \right| =$$

$$=\sqrt{\begin{matrix}[\varphi_0(1+2\mu_1)-5{,}8411]^2+[\varphi_0(1+\mu_1)-3{,}3177]^2+\\+[\varphi_0(1+3{,}5\mu_1)-9{,}6262]^2\end{matrix}}=A_2(\mu_1) \quad (60)$$

6) $\vec{z}_1(\vec{z}_2-\vec{z}_2^{k_2})=\varphi_0(\mu_1)[(1+2\mu_1)^2+(1+\mu_1)^2+(1+3{,}5\mu_1)^2]-$

$$-[18{,}785+48{,}6916\mu_1]=A_4(\mu_1) \quad (61)$$

7) $\left|\vec{a}_4(1)-\vec{z}_2^{k_2}\right|=$

$$=\left|a_{41}\vec{i}_1+(1{,}3707+\frac{1}{3}a_{41}\vec{i}_2+(-3{,}5163+2{,}25a_{41})\vec{i}_3-\right.$$

$$\left.-5{,}8411\vec{i}_1-3{,}3177\vec{i}_2-9{,}6262\vec{i}_3\right|=$$

$$=\sqrt{\begin{matrix}(a_{41}(1)-5{,}8411)^2+(-1{,}947+\frac{1}{3}a_{41}(1))^2+\\+(-13{,}1425+2{,}25a_{41}(1))^2\end{matrix}}=A_3(\mu_1)$$

8) $[\vec{a}_4(1)-\vec{z}_2^{k_2}]^2=$

$$=(a_{41}(1)-5{,}8411)^2+(-1{,}947+\frac{1}{3}a_{41}(1))^2+ \quad (63)$$

$$+(-13{,}1425+2{,}25a_{41}(1))^2$$

9) $(\vec{a}_3-\vec{z}_1^{k_1})(\vec{a}_4(1)-\vec{z}_2^{k_2})=$

$$=2(a_{41}(1)-6{,}1013)+1{,}5(-1{,}947+\frac{1}{3}a_{41}a_{41}(1))+ \quad (64)$$

$$+0{,}75(-13{,}1425+2{,}25a_{41}(1))$$

Substituting the values $a_{41}(1)$=6.5 and Eq. (54) in Eqs. (56)–(64), we have:

$$\left|\vec{a}_4(1)-\vec{z}_2^{k_2}\right|=1{,}9929 \tag{65}$$

$$[\vec{a}_4(1)-\vec{z}_2^{k_2}]^2=3{,}9715 \tag{66}$$

$$(\vec{a}_3-\vec{z}_1^{k_1})(\vec{a}_4(1)-\vec{z}_2^{k_2})=2{,}5532 \tag{67}$$

Now set up numerical relation between the parameters μ_3 and μ_1. For that, substituting Eqs. (65)–(67), and taking into account the numerical values $a_{41}(1) = 6{,}5$ and $\mu_2^{k_2}=2$, the relation Eq. (26) between the parameters will be of the form:

$$\mu_3=(\mu_2-2)\frac{2{,}5532}{3{,}9715} \text{ for } \mu_2 \geq 2,\ \mu_3 \geq 0$$

or

$$\mu_3=1{,}6429(\mu_2-2) \tag{68}$$

Substituting the numerical dependence between the parameters μ_2 and μ_1 in the form Eq. (42) in (68), set up dependence of the parameter μ_3 on the parameter μ_1 in the form:

$$\mu_3=0{,}7668(\mu_1-3{,}1768) \text{ for } \mu_1 \geq 3{,}1768 \tag{68a}$$

or

$$\mu_1=1{,}3041\cdot\mu_3+3{,}1768 \text{ for } \mu_3 \geq 0$$

Now, substituting Eqs. (56)–(58), (60)–(62), and (68) in Eq. (25), define the unaccounted factors predicting parameter λ_3 in the form:

$$\lambda_3=-0{,}0296\frac{\mu_1-3{,}1768}{\mu_1-1{,}5}\cdot\frac{A_1(\mu_1)A_2(\mu_1)A_3(\mu_1)}{A_4(\mu_1)}$$

for

$$\mu_1 \geq 3{,}1768 \tag{69}$$

where $\varphi_0(\mu_1)$ is of the form Eq. (47).

Now, by Eq. (31), calculate the cosine of the angle $cos\ \alpha_{23}$ between the economic process predicting vector function $\vec{Z}_3(1)$ and the second piecewise-linear vector-function $\vec{z}_2(\mu_2)$ in the form (Fig. 4.):

$$cos\alpha_{2,3} = \frac{(\vec{a}_4(1) - \vec{z}_2^{\,k_2})(\vec{z}_2(\mu_2) - \vec{z}_2^{\,k_2})}{\left|\vec{a}_4(1) - \vec{z}_2^{\,k_2}\right|\left|\vec{z}_2(\mu_2) - \vec{z}_2^{\,k_2}\right|} \tag{70}$$

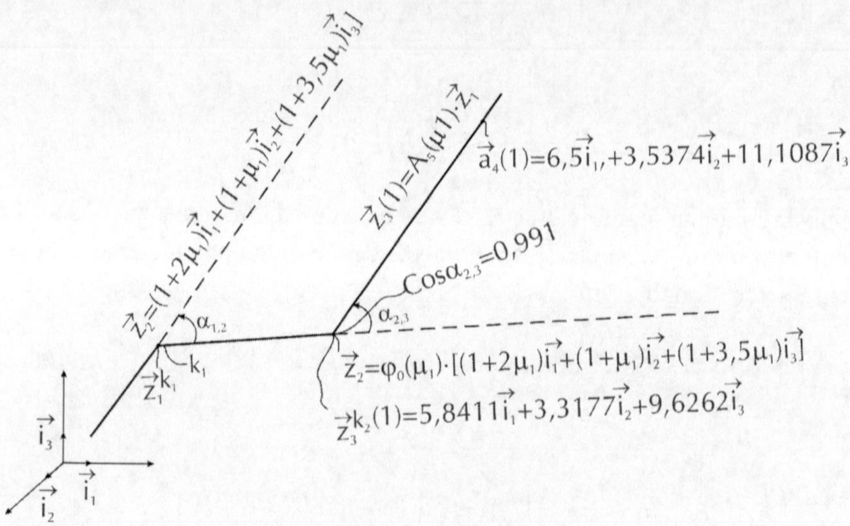

FIGURE 4 Numerical construction of predicting vector function $\vec{z}_3(\beta)$ on the base of 2-component economic-mathematical model in 3-dimensional vector space R_3.

Taking into account Eqs. (59)–(63), expression of $cos\ \alpha_{23}$ takes the form:
$cos\ \alpha_{23}=$

$$\cos\alpha_{2,3} =$$

$$= \frac{(6{,}7263\mu_1 + 2{,}3611)\varphi_0(\mu_1) - 18{,}8484}{1{,}6372\sqrt{\begin{array}{l}[\varphi_0(1+2\mu_1) - 5{,}8411]^2 + [\varphi_0(1+\mu_1) - 3{,}3177]^2 + \\ + [\varphi_0(1+3{,}5\mu_1) - 9{,}6262]^2\end{array}}}$$

For $\mu_1 = 5$ the numerical value of $\cos\,\alpha_{23}$ will be:

$$\cos\alpha_{2,3} = 0{,}9448 \tag{71}$$

From Eq. (24) calculate $\Omega_3(\lambda_3, \alpha_{2,3})$. For that substitute Eqs. (56)–(58), (60), (62), (68), (71) in Eq. (24), and calculate $\Omega_3(\lambda_3, \alpha_{2,3})$:

$$\Omega_3(\lambda_3, \alpha_{2,3}) = 0{,}028\frac{\mu_1 - 3{,}1768}{1{,}5 - \mu_1} \cdot \frac{A_1(\mu_1)A_2(\mu_1)A_3(\mu_1)}{A_4},$$

For

$$\mu_1 \geq 3{,}1768 \tag{72}$$

Now calculate the unaccounted parameter function $\omega_2(\lambda_2^{k_2}, \alpha_{1,2})$ belonging to the second piecewise-linear straight line, and take into account the character of relation between the parameters μ_2 and μ_1 given in the form Eq. (42):

$$\mu_2 = 1{,}1927(\mu_1 - 1{,}5) \text{ for } \mu_1 \geq 1{,}5,\ 0 \leq \mu_2 \leq \mu_2^{k_2} > 1 \tag{73}$$

Hence:

$$\mu_1 = 1{,}5 + 0{,}8384\mu_2 \tag{74}$$

For $\mu_2 = \mu_2^{k_2}$ from Eq. (74):

$$\mu_1^{k_2} = 1{,}5 + 0{,}8384\mu_2^{k_2} \tag{75}$$

For the considered example, for the second intersection point k_2 the value of the parameter $\mu_2^{k_2}$ earlier was accepted to be equal to 2, i.e., $\mu_2^{k_2} = 2$. In this case, the appropriate numerical value of the parameter $\mu_1^{k_2}$ by Eq. (75) will equal:

$$\mu_1^{k_2} = 3{,}1768 \tag{76}$$

Now carry out appropriate calculations by Eq. (21) for defining $\omega_2(\lambda_2^{k_2}, \alpha_{1,2})$, and calculate the vector $\vec{z}_1(\mu_1)$ in it for the value of the parameter $\mu_1 = \mu_1^{k_2} = 3{,}1768$. Taking into account $\mu_1^{k_1} = 1{,}5$, $\mu_2^{k_2} = 2$, $\mu_1^{k_2} = 3{,}1768$, cos $\alpha_{1,2} = 0{,}8495$, and also Eqs. (45), (56)–(58), define the numerical value of $\omega_2(\lambda_2^{k_2}, \alpha_{1,2})$ in the form:

$$\omega_2(\lambda_2^{k_2}, \alpha_{1,2}) = \lambda_2^{k_2} \cdot cos\,\alpha_{12} = -0{,}635 \tag{77}$$

Substituting Eqs. (56)–(58) in Eq. (23), express the coefficient A by the parameter $\mu_1 \geq \mu_1^{k_2} = 3{,}1768$ in the form:

$$A = -25{,}875\frac{\mu_1 - 1{,}5}{A_1(\mu_1)} \quad \text{for } \mu_1 \geq \mu_1^{k_2} = 3{,}1768 \tag{77a}$$

where

$$A_1(\mu_1) = 9{,}75 + 25{,}875\mu_1$$

Substituting the numerical values of the coefficient A Eq. (77a), the unaccounted parameter influence function $\omega_2(\lambda_2^{k_2}, \alpha_{1,2})$ Eq. (77) and also the unaccounted parameter influence predicting function $\Omega_3(\lambda_3, \alpha_{2,3})$ Eq. (72) in Eq. (20), for the case of 2-component piecewise-linear straight line find the form of the economic process predicting vector function $\vec{Z}_3(1)$ in 3-dimensional vector space in the form (Fig. 4) [4–6]:

$$\vec{Z}_3(1) = \vec{z}_1 \left\{ 1 - 9,4444 \frac{\mu_1 - 1,5}{A_1(\mu_1)} \cdot [1 - \right.$$

$$\left. - 0,0767 \frac{\mu_1 - 3,1768}{\mu_1 - 1,5} \cdot \frac{A_1(\mu_1) A_2(\mu_1) A_3(\mu_1)}{A_4(\mu_1)} \right\}$$

for

$$\mu_1 \geq 3,1768, \tag{78}$$

where

$$\vec{z}_1 = (1 + 2\mu_1)\vec{i}_1 + (1 + \mu_1)\vec{i}_2 + (1 + 3,5\mu_1)\vec{i}_3 \tag{79}$$

$$A_1\ (\mu_1) = 9.75 + 25,875\mu_1 \tag{80}$$

$$A_2(\mu_1) = \sqrt{\begin{matrix}[\varphi_0(1+2\mu_1) - 5,8411]^2 + [\varphi_0(1+\mu_1) - 3,3177]^2 + \\ + [\varphi_0(1+3,5\mu_1) - 9,6262]^2\end{matrix}} \tag{81}$$

$$A_3(\mu_1) = \sqrt{\begin{matrix}(a_{41} - 5,8411)^2 + (-1,947 + \frac{1}{3}a_{41})^2 + \\ + (-13,1425 + 2,25a_{41})^2\end{matrix}} \tag{82}$$

$$\begin{aligned} A_4(\mu_1) = {} & \varphi_0(\mu_1)[(1+2\mu_1)^2 + (1+\mu_1)^2 + (1+3,5\mu_1)^2] - \\ & - [18,785 + 48,6916\mu_1] \end{aligned} \tag{83}$$

$$\varphi_0(\mu_1) = 1 - (\mu_1 - 1{,}5) \cdot \frac{25{,}8751}{9{,}75 + 25{,}875 \cdot \mu_1} \left[1 + \right.$$

$$\left. + 0{,}1026 \frac{9{,}75 + 25{,}875 \cdot \mu_1}{9{,}75 + 19{,}375 \cdot \mu_1 - 17{,}25 \mu_1^2} \sqrt{38{,}8125 - 51{,}75 \cdot \mu_1 + 17{,}25 \mu_1^2} \right] \tag{84}$$

Eq. (78) is written in the compact form as follows (Fig. 5):

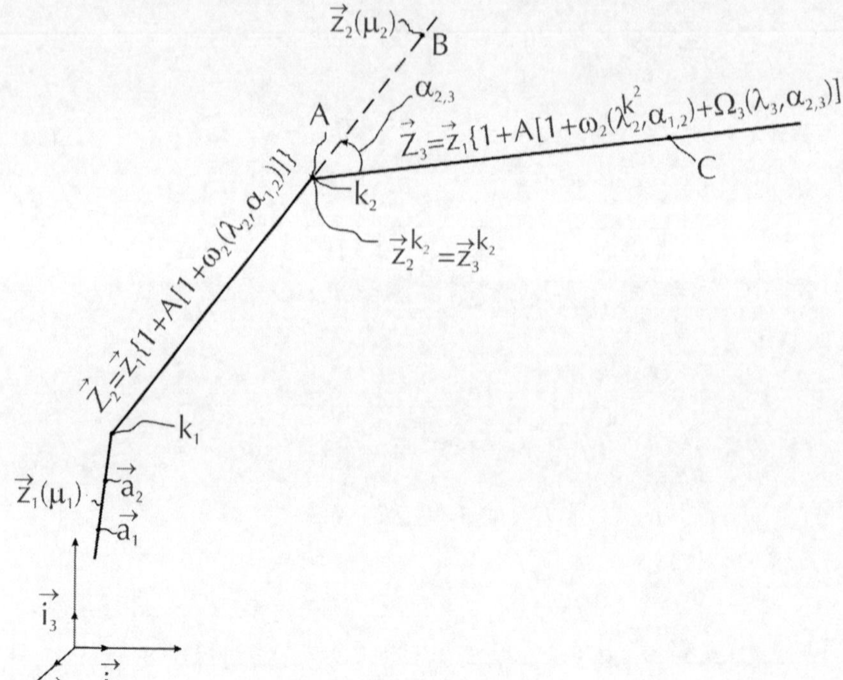

FIGURE 5 Compact form of representation of numerical expression of the predicting vector function $\vec{Z}_3(\beta)$ constructed on the base of 2-component model in 3-dimensional vector space R_3.

$$\vec{Z}_3(1) = A_5(\mu_1) \cdot \vec{z}_1 \quad (85)$$

where

$$A_5(\mu_1) = 1 - 9{,}4444 \frac{\mu_1 - 1{,}5}{A_1(\mu_1)} \cdot [1 -$$

$$-0{,}0767 \frac{\mu_1 - 3{,}1768}{\mu_1 - 1{,}5} \cdot \frac{A_1(\mu_1) A_2(\mu_1) A_3(\mu_1)}{A_4(\mu_1)}$$

for

$$\mu_1 \geq 3{,}1768 \quad (86)$$

3.2.3 COMPONENT PIECEWISE-LINEAR ECONOMIC-MATHEMATICAL MODE AND METHOD OF MULTIVARIANT PREDICTION OF ECONOMIC PROCESS WITH REGARD TO UNACCOUNTED FACTORS INFLUENCE IN 3-DIMENSIONAL VECTOR SPACE

In this section, we have given numerical construction of 3-component piecewise-linear economic model with regard to unaccounted factors influence in 3-dimensional vector space, and construct appropriate vector-functions on the subsequent ΔV_4 small volume of 3-dimensional space [4–6, 14, 15].

Consider the case when the economic process given in the form of the statistical points (vectors) set $\{\vec{a}_n\}$ in 3-dimensional vector space R_3 is represented in the form of 3 piecewise-linear straight lines of the form Eq. (78) of Chapter 2:

$$\vec{z}_1 = \vec{a}_1 + \mu_1(\vec{a}_2 - \vec{a}_1) \tag{87}$$

$$\vec{z}_2 = \vec{a}_2 + \mu_2(\vec{a}_3 - \vec{a}_2) \tag{88}$$

$$\vec{z}_3 = \vec{a}_3 + \mu_3(\vec{a}_4 - \vec{a}_3) \tag{89}$$

Here the vectors $\vec{z}_1 = \vec{z}_1(z_{11}, z_{12}, z_{13})$, $\vec{z}_2 = \vec{z}_2(z_{21}, z_{22}, z_{23})$ and $\vec{z}_3 = \vec{z}_3(z_{31}, z_{32}, z_{33})$ with the coordinates z_{ij} are given in the form of linear vector functions for the first, second and third piecewise-linear straight lines in 3-dimensional vector-space; the vectors (points) $\vec{a}_1(a_{11}, a_{12}, a_{13})$, $\vec{a}_2 = \vec{a}_2(a_{21}, a_{22}, a_{23})$, $\vec{a}_3 = \vec{a}_3(a_{31}, a_{32}, a_{33})$ and $\vec{a}_4 = \vec{a}_4(a_{41}, a_{42}, a_{43})$ are the given of 3-dimensional vector space R_3; $\mu_1 \geqslant 0$, $\mu_2 \geqslant 0$, and $\mu_3 \geq 0$ are arbitrary parameters of the first, second and third piecewise-linear straight lines. It holds the equality $\lambda_1 + \mu_1 = 1$, $\lambda_2 + \mu_2 = 1$ and $\lambda_3 + \mu_3 = 1$; $\alpha_{1,2}$ and $\alpha_{2,3}$ are the adjacent angles between the first and second and also between the second and third piecewise-linear straight lines; k_1 and k_2 are the intersection points between the piecewise-linear straight lines (Fig. 5).

As the intersection point between the second and third piecewise-linear straight lines in 3-dimensional vector space may also not coincide with the point $\vec{a}_3$, we denote this intersection point by $\vec{z}_2^{k_2}$ (Fig. 5). With regard to this factor, according to Eq. (45) of Chapter 2, an equation for the third piecewise-linear straight line is written in the form:

$$\vec{z}_3 = \vec{z}_2^{k_2} + \mu_3(\vec{a}_4 - \vec{z}_2^{k_2}) \tag{90}$$

Here $\vec{z}_2^{k_2}$ is the value of the point (vector) of the second piecewise-linear straight line at the k_2-th intersection point, represented by Eq. (40) of Chapter 2 and calculated for the value of the parameter $\mu_1 \geq \mu_1^{k_2}$, i.e., at the second intersection point k_2 and equal:

$$\vec{z}_2^{k_2} = \vec{z}_1^{k_2}[1 - A(\mu_1^{k_2})(1 + \omega_2(\lambda_2^{k_2}, \alpha_{1,2})]\} \tag{91}$$

where the parameter $\mu_1^{k_2}$ is calculated by Eq. (34a) of Chapter 2:

$$\mu_1^{k_2} = \mu_1^{k_1} + \mu_2^{k_2} \frac{(\vec{a}_3 - \vec{z}_1^{k_1})^2}{(\vec{a}_3 - \vec{z}_1^{k_1})(\vec{a}_2 - \vec{a}_1)} \tag{92}$$

and the vector $\vec{z}_1^{k_2}$ is calculated by means of Eq. (87) at the point $\mu_1 = \mu_1^{k_2}$ in the form:

$$\vec{z}_1^{k_2} = \vec{a}_1 + \mu_1^{k_2}(\vec{a}_2 - \vec{a}_1) \tag{93}$$

and the coefficients $A(\mu_1^{k_2})$ and $\omega_2(\lambda_2^{k_2}, \alpha_{1,2})$ are calculated by Eqs. (25), and (28) of Chapter 2, and Eq. (11) at the point $\mu_1 = \mu_1^{k_2}$ in the form:

$$A(\mu_1^{k_2}) = (\mu_1^{k_1} - \mu_1^{k_2}) \frac{|\vec{a}_2 - \vec{a}_1||\vec{z}_1^{k_1} - \vec{a}_1|}{\vec{z}_1(\mu_1^{k_2})(\vec{z}_1^{k_1} - \vec{a}_1)} \tag{94}$$

$$\lambda_2^{k_2} = \frac{\mu_2^{k_2}}{\mu_1^{k_1} - \mu_1^{k_2}} \cdot \frac{|\vec{z}_1(\mu_1^{k_2}) - \vec{z}_1^{k_1}||\vec{a}_3 - \vec{z}_1^{k_1}|}{\vec{z}_1(\mu_1^{k_2})(\vec{z}_1(\mu_1^{k_2}) - \vec{z}_1^{k_1})} \frac{\vec{z}_1(\mu_1^{k_2})(\vec{z}_1^{k_1} - \vec{a}_1)}{|\vec{a}_2 - \vec{a}_1||\vec{z}_1^{k_1} - \vec{a}_1|} \tag{95}$$

$$cos\alpha_{1,2} = \frac{(\vec{z}_1(\mu_1^{k_2}) - \vec{z}_1^{k_1})(\vec{a}_3 - \vec{z}_1^{k_1})}{|\vec{z}_1(\mu_1^{k_2}) - \vec{z}_1^{k_1}| \cdot |\vec{a}_3 - \vec{z}_1^{k_1}|} \tag{96}$$

$$\omega_2(\lambda_2^{k_2}, \alpha_{1,2}) = \lambda_2^{k_2} cos\alpha_{1,2} \tag{97}$$

Here the value of the parameter at the second intersection point $\mu_2^{k_2}$ corresponding to the final point of the second piecewise-linear straight line is connected with the appropriate value of the parameter $\mu_1^{k_2}$ acting on the first straight line in the form Eq. (92):

$$\mu_2^{k_2}=(\mu_1^{k_2}-\mu_1^{k_1})\frac{(\vec{a}_3-\vec{z}_1^{k_1})(\vec{a}_2-\vec{a}_1)}{(\vec{a}_3-\vec{z}_1^{k_1})^2} \qquad (98)$$

Thus, giving the values of the parameter $\mu_1^{k_2}$ and $\mu_2^{k_2}$ at the intersection point k_1 and k_2 by Eq. (96) or Eq. (98), it is easy to define the appropriate value of the parameter $\mu_1^{k_2}$.

Using Eqs. (70)–(74) of Chapter 2, in 3-dimensional vector space write an equation for the points of the third piecewise-linear straight line depending on the vector equation of the first piecewise-linear straight line, spatial form of unaccounted parameters $\lambda_2^{k_2}$ and λ_3, and also on unaccounted parameters spatial influence functions $\omega_2(\lambda_2^{k_2},\alpha_{1,2})$ and $\omega_3(\lambda_3,\alpha_{2,3})$ in the form (Fig. 5):

$$\vec{z}_3=\vec{z}_1\{1+A\ [1+\omega_2(\lambda_2^{k_2},\alpha_{1.2})+\omega_3(\lambda_3,\alpha_{2,3})]\}$$

for

$$\mu_1 \geq \mu_1^{k_2},\ \mu_2 \geq \mu_2^{k_2} \qquad (99)$$

Here the unaccounted parameter influence function $\omega_2(\lambda_2^{k_2},\alpha_{1,2})$ is calculated by means of Eq. (11) of Chapter 2, the unaccounted parameter influence function $\omega_3(\lambda_3,\alpha_{2,3})$ is calculated in the form Eqs. (73)–(76) of Chapter 2:

$$\omega_3(\lambda_3,\alpha_{2,3})=\lambda_3\cdot cos\alpha_{2,3} \text{ for } \mu_3\geq 0,\quad \mu_2\geq\mu_2^{k_2}$$

$$\lambda_3=\frac{\mu_3}{\mu_1^{k_1}-\mu_1}\cdot\frac{\left|\vec{z}_2-\vec{z}_2^{k_2}\right|\left|\vec{a}_4-\vec{z}_2^{k_2}\right|}{\vec{z}_1(\vec{z}_2-\vec{z}_2^{k_2})}\cdot\frac{\vec{z}_1(\vec{z}_1^{k_1}-\vec{a}_1)}{\left|\vec{a}_2-\vec{a}_1\right|\left|\vec{z}_1^{k_1}-\vec{a}_1\right|}$$

for

$$\mu_2\geq\mu_2^{k_2},\ \mu_3\geq 0 \qquad (100)$$

$$cos\alpha_{2,3} = \frac{(\vec{z}_2(\mu_2) - \vec{z}_2^{k_2})(\vec{a}_4 - \vec{z}_2^{k_2})}{\left|\vec{z}_2(\mu_2) - \vec{z}_2^{k_2}\right|\left|\vec{a}_4 - \vec{z}_2^{k_2}\right|}, \text{ for } \mu_2 \geq \mu_2^{k_2} \tag{101}$$

$$\mu_3 = (\mu_2 - \mu_2^{k_2})\frac{(\vec{a}_3 - \vec{z}_1^{k_1})(\vec{a}_4 - \vec{z}_2^{k_2})}{(\vec{a}_4 - \vec{z}_2^{k_2})^2} \text{ for } \mu_2 \geq \mu_2^{k_2} \tag{102}$$

$$\mu_2 = (\mu_1 - \mu_1^{k_1})\frac{(\vec{a}_3 - \vec{z}_1^{k_1})(\vec{a}_2 - \vec{a}_1)}{(\vec{a}_3 - \vec{z}_1^{k_1})^2} \text{ for } \mu_2 \geqslant \mu_1^{k_1} \tag{103}$$

$$\mu_2^{k_2} = (\mu_1^{k_2} - \mu_1^{k_1})\frac{(\vec{a}_3 - \vec{z}_1^{k_1})(\vec{a}_2 - \vec{a}_1)}{(\vec{a}_3 - \vec{z}_1^{k_1})^2} \text{ for } \mu_1^{k_2} \geqslant \mu_1^{k_1} \tag{104}$$

$$A = (\mu_1^{k_1} - \mu_1)\frac{\left|\vec{a}_2 - \vec{a}_1\right|\left|\vec{z}_1^{k_1} - \vec{a}_1\right|}{\vec{z}_1\,(\vec{z}_1^{k_1} - \vec{a}_1)} \tag{105}$$

Thus, giving the vectors (point) $\vec{a}_1, \vec{a}_2, \vec{a}_3, \vec{a}_4$, $\vec{z}_1^{k_1}$, $\vec{z}_2^{k_2}$ and $\vec{z}_2(\mu_2)$ Eq. (99) will represent a vector equation for the third piecewise-linear straight line $\vec{z}_3 = \vec{z}_3(\mu_1, \omega_3)$ in 3-dimensional vector space depending on the parameter $\mu_1 \geqslant \mu_1^{k_2}$ (i.e., for $\mu_2 \geqslant \mu_2^{k_2}$) and unaccounted parameters influence functions $\omega_2(\lambda_2^{k_2}, \alpha_{1,2})$ and $\omega_3(\lambda_3, \alpha_{2,3})$.

Note that Eq. (99) defines all the points of the third piecewise-linear straight line in 3-dimensional space. To the case $\mu_3 = 0$ there will correspond the value of the initial point of the third straight line that will be expressed by the vector-function of the first piecewise-linear straight line $\vec{z}_1$, by the value of the parameters of intersection points of piecewise-linear straight lines $\mu_1^{k_1}$ and $\mu_2^{k_2}$, and also $Cos\alpha_{1,2}$ generated between the first and second piecewise-linear straight lines. It will equal:

$$\vec{z}_3\Big|_{\mu_3=0} = \vec{z}_1\left\{1+(\mu_1^{k_1}-\mu_1)\frac{|\vec{a}_2-\vec{a}_1||\vec{z}_1^{k_1}-\vec{a}_1|}{\vec{z}_1(\vec{z}_1^{k_1}-\vec{a}_1)}\cdot\left[1+\right.\right.$$

$$\left.\left.+\frac{\mu_2^{k_2}}{\mu_1^{k_1}-\mu_1}\frac{|\vec{z}_1-\vec{z}_1^{k_1}||\vec{a}_3-\vec{z}_1^{k_1}|}{\vec{z}_1(\vec{z}_1-\vec{z}_1^{k_1})}\cdot\frac{\vec{z}_1(\vec{z}_1^{k_1}-\vec{a}_1)}{|\vec{a}_2-\vec{a}_1||\vec{z}_1^{k_1}-\vec{a}_1|}\tilde{n}os\alpha_{1,2}\right]\right\} \tag{106}$$

Write the coordinate form of vector equation Eq. (99). Therefore, we have to take into account that in 3-dimensional vector space $\vec{z}_3 = \sum_{m=1}^{3} z_{3m}\,\vec{i}_m$ and $\vec{z}_1 = \sum_{m=1}^{3} z_{1m}\,\vec{i}_m$. In this case, the coordinates of the vector $\vec{z}_3$ i.e., z_{3m}, will be expressed by the coordinates of the first piece-wise-linear straight line z_{1m}, spatial form of unaccounted parameters $\lambda_2^{k_2}$ and λ_3, and also on unaccounted parameters influence functions $\omega_2(\lambda_2^{k_2},\alpha_{1,2})$ and $\omega_3(\lambda_3,\alpha_{2,3})$ in the form:

$$z_3 = z_{1m}\{1+A\,[1+\omega_2(\lambda_2^{k_2},\alpha_{1,2})+\omega_3(\lambda_3,\alpha_{2,3})]\},\ (m=1,2,3) \tag{107}$$

Here

$$A=(\mu_1^{k_1}-\mu_1)\frac{\sum_{i=1}^{3}(a_{2i}-a_{1i})^2}{\sum_{i=1}^{3}(a_{2i}-a_{1i})[a_1+\mu_1(a_{2i}-a_{1i})]} \tag{108}$$

$$\lambda_2^{k_2}=\frac{\mu_2^{k_2}}{\mu_1^{k_1}-\mu_1^{k_2}}\cdot\frac{\sqrt{\sum_{i=1}^{3}\{a_{3i}-[a_{1i}+\mu_1^{k_1}(a_{2i}-a_{1i})]\}^2}}{\sqrt{\sum_{i=1}^{3}(a_{2i}-a_{1i})^2}} \tag{109}$$

$$\omega_2(\lambda_2^{k_2}, \alpha_{1,2}) = \lambda_2^{k_2} \cdot cos\alpha_{1,2} \tag{110}$$

$$\omega_3(\lambda_3, \alpha_{2,3}) = \lambda_3 \cdot cos\alpha_{2,3} =$$

$$= \frac{\mu_3}{\mu_1^{k_1} - \mu_1} \cdot \frac{\sqrt{\sum_{i=1}^{3}(z_{2i} - z_{2i}^{k_2})^2} \cdot \sqrt{\sum_{i=1}^{3}(a_{4i} - z_{2i}^{k_2})^2}}{\sqrt{\sum_{i=1}^{3}(z_{2i} - z_{2i}^{k_2})}} \cdot$$

$$\cdot \frac{\sum_{i=1}^{3} z_{1i}(a_{2i} - a_{1i})}{\sum_{i=1}^{3}(a_{2i} - a_{1i})^2} \cdot cos\alpha_{2,3} \tag{111}$$

$$\mu_3 = (\mu_2 - \mu_2^{k_2}) \frac{\sum_{i=1}^{3}(a_{3i} - z_{1i}^{k_1})(a_{4i} - z_{2i}^{k_2})}{\sum_{i=1}^{3}(a_{4i} - z_{2i}^{k_2})^2}, \text{ for } \mu_2 \geq \mu_2^{k_2} \tag{112}$$

$$\mu_2 = (\mu_1 - \mu_1^{k_1}) \frac{\sum_{i=1}^{3}(a_{3i} - z_{1i}^{k_1})(a_{2i} - a_{1i})}{\sum_{i=1}^{3}(a_{3i} - z_{1i}^{k_1})^2}, \text{ for } \mu_1 \geqslant \mu_1^{k_1} \tag{113}$$

$$\mu_2^{k_2} = (\mu_1^{k_2} - \mu_1^{k_1}) \frac{\sum_{i=1}^{3}(a_{3i} - z_{1i}^{k_1})(a_{2i} - a_{1i})}{\sum_{i=1}^{3}(a_{3i} - z_{1i}^{k_1})^2}, \text{ for } \mu_1^{k_2} \geqslant \mu_1^{k_1} \tag{114}$$

Now for the case of economic process represented in the form of 3-component piecewise-linear economic-mathematical model investigate the prediction and control of such a process on the subsequent $\Delta V_4(x_1,x_2,x_3)$ small volume of 3-dimensional vector space with regard to unaccounted parameters influence functions $\omega_2(\lambda_2^{k_2},\alpha_{1,2})$ and $\omega_3(\lambda_3^{k_3},\alpha_{2,3})$. The values of the unaccounted parameters functions $\omega_2(\lambda_2,\alpha_{1,2})$ and $\omega_3(\lambda_3,\alpha_{2,3})$ are unknown [4–6, 14].

In Section 2.3, we developed a method for constructing an economic process predicting vector function $\vec{Z}_{N+1}(\beta)$ with regard to the introduced unaccounted parameter influence predicting function $\Omega_{N+1}(\lambda_{N+1},\alpha_{N,N+1})$ in the m-th vector space that found its reflection in Eqs. (105)–(111). Apply this method to the case of the given 3-component piecewise-linear economic process in 3-dimensional vector space. In this case, the predicting vector function $\vec{Z}_4(\beta)$ will be of the form (Fig. 6):

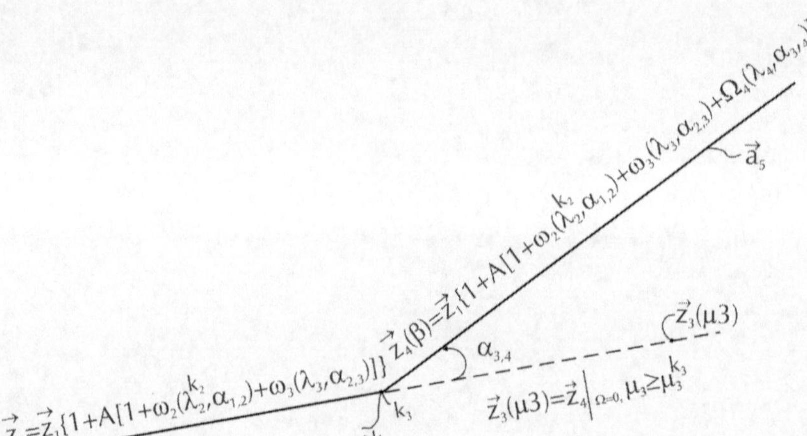

FIGURE 6 Construction of predicting vector function $\vec{Z}_4(\beta)$ with regard to unaccounted parameter influence predicting function $\Omega_4(\lambda_4,\alpha_{3,4})$ on the base of 3-component piecewise-linear economic-mathematical model in 3-dimensional vector space R_3.

$$\vec{Z}_4(\beta)=\vec{z}_1\{1+A[1+\omega_2(\lambda_2^{k_2},\alpha_{1,2})+\omega_3(\lambda_3^{k_3},\alpha_{2,3})+\Omega_4(\lambda_4,\alpha_{3,4})]\}$$

for $$\beta = 1,2 \tag{115}$$

where

$$\Omega_4(\lambda_4, \alpha_{3,4}) = \lambda_4 \cdot cos\alpha_{3,4} \tag{116}$$

$$\lambda_4 = \frac{\mu_4}{\mu_1^{k_1} - \mu_1} \cdot \frac{\left|\vec{z}_3 - \vec{z}_3^{k_3}\right| \cdot \left|\vec{a}_5(\beta) - \vec{z}_3^{k_3}\right|}{\vec{z}_1(\vec{z}_3 - \vec{z}_3^{k_3})} \cdot \frac{\vec{z}_1(\vec{z}_1^{k_1} - \vec{a}_1)}{\left|\vec{a}_2 - \vec{a}_1\right| \cdot \left|\vec{z}_1^{k_1} - \vec{a}_1\right|}$$

for

$$\beta = 1,2, \ \mu_1 \geqslant \mu_3^{k_3}, \qquad \mu_4 \ge 0 \tag{117}$$

$$\mu_4 = (\mu_3 - \mu_3^{k_3}) \cdot \frac{(\vec{a}_4 - \vec{z}_2^{k_2})(\vec{a}_5(\beta) - \vec{z}_3^{k_3})}{(\vec{a}_5(\beta) - \vec{z}_3^{k_3})^2},$$

$$\mu_3 \ge \mu_3^{k_3}, \ \mu_4 \ge 0 \text{ for } \beta = 1,2 \tag{118}$$

The expressions of the unaccounted parameters functions $\omega_2(\lambda_2^{k_2}, \alpha_{1,2})$ and $\omega_3(\lambda_3^{k_3}, \alpha_{2,3})$ have the form Eq. (107)–(109) and

$$\omega_2(\lambda_2^{k_2}, \alpha_{1,2}) = \lambda_2^{k_2} \cdot cos\alpha_{1,2} =$$

$$= \frac{\mu_2^{k_2}}{\mu_1^{K_1} - \mu_1} \cdot \frac{\left|\vec{z}_1 - \vec{z}_1^{k_1}\right| \cdot \left|\vec{a}_3 - \vec{z}_1^{k_1}\right|}{\vec{z}_1(\vec{z}_1 - \vec{z}_1^{k_1})} \cdot \frac{\vec{z}_1(\vec{z}_1^{k_1} - \vec{a}_1)}{\left|\vec{a}_2 - \vec{a}_1\right| \cdot \left|\vec{z}_1^{k_1} - \vec{a}_1\right|} cos\alpha_{1,2} \tag{119}$$

$$\mu_2 = (\mu_1 - \mu_1^{k_1}) \cdot \frac{(\vec{a}_2 - \vec{a}_1)(\vec{a}_3 - \vec{z}_1^{k_1})}{(\vec{a}_3 - \vec{z}_1^{k_1})^2}, \qquad \mu_1 \ge \mu_1^{k_1} \tag{120}$$

$$A = (\mu_1^{k_1} - \mu_1)\cdot\frac{\left|\vec{a}_2 - \vec{a}_1\right|\cdot\left|\vec{z}_1^{k_1} - \vec{a}_1\right|}{\vec{z}_1(\vec{z}_1^{k_1} - \vec{a}_1)} \tag{121}$$

Here the vector $\vec{a}_5(\beta)$ for each value of $\beta = 1,2$, according Eq. (18) of Chapter 2 is of the form:

$$\vec{a}_5(\beta) = a_{51}(\beta)\vec{i}_1 + a_{52}(\beta)\vec{i}_2 + a_{53}(\beta)\vec{i}_3 = \sum_{m=1}^{3} a_{5m}(\beta)\cdot\vec{i}_m$$

for

$$\beta = 1,2 \tag{122}$$

And by means of Eq. (91) of Chapter 2, the coordinates of the vector $\vec{a}(\beta)$ will be expressed by the coordinates of the vectors $\vec{a}_{\beta+1}$, $\vec{z}_{\beta\text{-}1}^{k_{\beta\text{-}1}}$ and $\vec{z}_3^{k_3}$ in the form:

$$C_\beta = \frac{a_{51}(\beta) - z_{31}^{k_3}}{a_{\beta+1,1} - z_{\beta\ 1,1}^{k_{\beta\ 1}}} = \frac{a_{52}(\beta) - z_{32}^{k_3}}{a_{\beta+1,2} - z_{\beta\ 1,2}^{k_{\beta\ 1}}} = \frac{a_{53}(\beta) - z_{33}^{k_3}}{a_{\beta+1,3} - z_{\beta\ 1,3}^{k_{\beta\ 1}}} \tag{123}$$

Hence, by Eq. (123), the coordinates $a_{52}(\beta)$ and $a_{53}(\beta)$ will be expressed by the arbitrarily given coordinate $a_{51}(\beta) > z_{31}^{k_3}$, in the form:

$$a_{52}(\beta) = z_{32}^{k_3} + (a_{51}(\beta) - z_{31}^{k_3})\frac{a_{\beta+1,2} - z_{\beta\text{-}1,2}^{k_{\beta\text{-}1}}}{a_{\beta+1,1} - z_{\beta\text{-}1,1}^{k_{\beta\text{-}1}}}$$

$$(124)$$

$$a_{53}(\beta) = z_{33}^{k_3} + (a_{51}(\beta) - z_{31}^{k_3})\frac{a_{\beta+1,3} - z_{\beta\text{-}1,3}^{k_{\beta\text{-}1}}}{a_{\beta+1,1} - z_{\beta\text{-}1,1}^{k_{\beta\text{-}1}}}$$

Here the coefficients $a_{\beta+1,m}$ and $z_{\beta\text{-}1,m}^{k_{\beta\text{-}1}}$ are the coordinates of the vectors $\vec{a}_{\beta\text{-}1}$ and $\vec{z}_{\beta\text{-}1}$ in 3-dimensional vector space and equal:

$$\vec{a}_{\beta\text{-}1} = \sum_{m=1}^{3} a_{\beta\text{-}1,m} \vec{i}_m, \; \vec{z}_{\beta\text{-}1} = \sum_{m=1}^{3} z_{\beta\text{-}1,m} \vec{i}_m \tag{125}$$

Note that in the vectors $\vec{z}_4(\beta)$ and $\vec{a}_5(\beta)$ the index (β) in the parenthesis means that the vector $\vec{z}_4(\beta)$ is parallel to the β-th piecewise-linear vector-function $\vec{z}_\beta$. This will mean that the occurring economic process, beginning with the point $\vec{z}_3^{k_3}$ will occur by the scenario of the β-th piecewise-linear equation. In our example $\beta = 1,2$. In our case, there will be three predicting functions, i.e., $\vec{z}_4(1)$, $\vec{z}_4(2)$ and the case when the influence of unaccounted factors $\vec{z}_4(0)$ will not be available. In all these cases, the predicting vector-functions $\vec{z}_\beta(\beta)$ will emanate from one point $\vec{z}_3^{k_3}$, and the predicting vector-function $\vec{z}_4(1)$ will be parallel to the first piecewise-linear straight line; $\vec{z}_4(2)$ will be continuation of the third vector straight line $\vec{z}_4(0)$, and all of them will emanate from one point $\vec{z}_3^{k_3}$.

The expression $cos\alpha_{3,4}$ corresponding to the cosine of the angle between the third piecewise-linear straight line $\vec{z}_3$ and the predicting fourth vector straight line $\vec{z}_\beta(\beta)$ for each value of β on the base of scalar product of two vectors is represented in the form (Fig. 6):

$$cos\alpha_{3,4} = \frac{(\vec{z}_3 - \vec{z}_3^{k_3})(\vec{a}_5(\beta) - \vec{z}_3^{k_3})}{\left|\vec{z}_3 - \vec{z}_3^{k_3}\right|\left|\vec{a}_5(\beta) - \vec{z}_3^{k_3}\right|} \tag{127}$$

3.2.1 METHOD OF NUMERICAL CALCULATION OF 3-COMPONENT ECONOMIC-MATHEMATICAL MODEL AND DEFINITION OF PREDICTING VECTOR FUNCTION WITH REGARD TO UNACCOUNTED FACTORS INFLUENCE IN 3-DMENSIONAL VECTOR SPACE

In this section, we have given numerical calculates of 3-component piecewise-linear economic-mathematical model with regard to unaccounted parameters influence in 3-dimensional vector space, and construct appropriate predicting vector functions $\vec{z}_4(1)$, $\vec{z}_4(2)$ and $\vec{z}_4(0)$ on subsequent ΔV_4 small volume of 3-dimensional space [3–6, 15, 16].

Consider the case of economic process given in the form of the statistical points (vectors) set $\{a_n\}$ in 3-dimensional vector space R_3 represented in the form of 3-component piecewise-linear function of the form Eqs. (87)–(89). The vectors $\vec{a}_i = \vec{a}_i(a_{i1}, a_{i2}, a_{i3})$ (where $i = 1,2,3$) are the given points of 3-dimensional vector space R_3 and have the form:

$$\vec{a}_1 = \vec{i}_1 + \vec{i}_2 + \vec{i}_3,$$

$$\vec{a}_2 = 3\vec{i}_1 + 2\vec{i}_2 + 4{,}5\vec{i}_3,$$

$$\vec{a}_3 = 6\vec{i}_1 + 4\vec{i}_2 + 7\vec{i}_3,$$

$$\vec{a}_4 = 8\vec{i}_1 + 9\vec{i}_2 + 10\vec{i}_3 \tag{127}$$

Below, by means of these vectors we have showed a method for calculating a chain form of each piecewise-linear vector equation depending on the first piecewise-linear vector straight line $\vec{z}_1$, cosines of the angles $cos\alpha_{1,2}$ and $cos\alpha_{2,3}$ generated between the adjacent first and second and also third and fourth piecewise-linear vector lines, and also on the parameter μ_1 corresponding to the first vector line [4–6, 17].

Substituting Eq. (127) in Eq. (87), the equation of the first straight line in the coordinate form will be of the form:

$$\vec{z}_1 = (1 + 2\mu_1)\vec{i}_1 + (1 + \mu_1)\vec{i}_1 + (1 + 3{,}5\mu_1)\vec{i}_3 \tag{128}$$

Giving the value of the parameter μ_1 for the intersection point k_1 between the first and second piecewise-linear straight lines of the form $\mu_1^{k_1} = 1{,}5$, the coordinate form of the intersection point $\vec{z}_1^{k_1}$ is defined from Eq. (128) in the form:

$$\vec{z}_1 = (1 + 2\mu_1)\vec{i}_1 + (1 + \mu_1)\vec{i}_1 + (1 + 3{,}5\mu_1)\vec{i}_3 \tag{129}$$

By means of intersection point Eq. (129) and the given point $\vec{a}_3$ on the second straight line, by Eq. (103) set up a numerical relation between the parameters μ_1 and μ_2 in the form:

$$\mu_2 = 1{,}1927(\mu_1 - 1{,}5),\ \text{for}\ \mu_1 \geqslant 1{,}5,\ \mu_2 \geqslant 0 \tag{130}$$

Hence:

$$\mu_1 = 1{,}5 + 0{,}8384\mu_2 \tag{131}$$

Eq. (130) means that on the second piecewise-linear straight line, to the value of the parameter μ_2 there will be determined appropriate value of the operator μ_1 by Eq. (131). For the given value of the parameter $\mu_2^{k_2}$, corresponding to the intersection point between the second and third piecewise-linear straight lines equal 2, i.e., $\mu_2^{k_2} = 2$ from Eq. (131) or Eq. (104) the appropriate value of the parameter $\mu_1^{k_2}$ will equal:

$$\mu_1^{k_2} = 3{,}1768 \tag{132}$$

This means that when the parameter μ_2 corresponding to the points of the second piecewise-linear straight line will change within $0 \leqslant \mu_2 \leqslant 2$, then the appropriate value of the parameter μ_1 will change in the interval:

$$1{,}5 \leqslant \mu_1 \leqslant 3{,}1768 \tag{133}$$

This case will correspond to the case of the segment of the second straight line. For the value of the parameter $\mu_2 \geqslant 2$ the appropriate value of the parameter μ_1, will be $\mu_1 \geqslant 3{,}1768$. This case will correspond to the vector equation of the second straight line restricted from one end.

Now establish the form of the vector equation of the second piecewise-linear straight line depending on the vector equation of the first piecewise-linear straight line $\vec{z}_1$, $\cos\alpha_{1,2}$ and the parameter μ_1. For that we use the constructed general Eq. (5):

$$\vec{z}_2 = \vec{z}_1\{1 + A[1 + \omega_2(\lambda_2, \alpha_{1,2})]\} \tag{134}$$

where the coefficient A, the unaccounted factor parameter λ_2 and the unaccounted parameters function $\omega_2(\lambda_2, \alpha_{1,2})$ will be of the form:

$$A = (\mu_1^{k_1} - \mu_1)\frac{\left|\vec{z}_1^{k_1} - \vec{a}_1\right|\left|\vec{a}_2 - \vec{a}_1\right|}{\vec{z}_1(\vec{z}_1^{k_1} - \vec{a}_1)} \tag{135}$$

$$\lambda_2 = \frac{\mu_2}{\mu_1^{k_1} - \mu_1}\frac{\left|\vec{z}_1 - \vec{z}_1^{k_1}\right|\left|\vec{a}_3 - \vec{z}_1^{k_1}\right|}{\vec{z}_1(\vec{z}_1 - \vec{z}_1^{k_1})}\frac{\vec{z}_1(\vec{z}_1^{k_1} - \vec{a}_1)}{\left|\vec{a}_2 - \vec{a}_1\right|\left|\vec{z}_1^{k_1} - \vec{a}_1\right|} \tag{136}$$

$$\omega_2(\lambda_2, \alpha_{1,2}) = \lambda_2 \cdot cos\alpha_{1,2} \tag{137}$$

$$cos\alpha_{1,2} = \frac{(\vec{z}_1(\mu_1^{k_2}) - \vec{z}_1^{k_1})(\vec{a}_3 - \vec{z}_1^{k_1})}{\left|\vec{z}_1(\mu_1^{k_2}) - \vec{z}_1^{k_1}\right|\left|\vec{a}_3 - \vec{z}_1^{k_1}\right|} \tag{138}$$

Note that by Eq. (134) we must carry out numerical calculation for the values of the parameter μ_1, changing in the interval represented by Eq. (140a). In conformity to our problem, we should use the range of the parameter μ_1 given in Eq. (133).

Determine the numerical values of the coefficients A, λ_2, $cos\alpha_{1,2}$ and $\omega_2(\lambda_2, \alpha_{1,2,})$. For that substitute Eqs. (127)–(130) and the value of the parameter $\mu_1^{k_1} = \mu_1^{1,5} = 1{,}5$ in Eq. (135)–(138), and get:

$$A = -25{,}875 \cdot \varphi_0(\mu_1), \text{ for } 1{,}5 \le \mu_1 \le 3{,}1768 \tag{139}$$

$$\lambda_2 = 0{,}1203 \cdot \varphi_1(\mu_1), \text{ for } 1{,}5 \le \mu_1 \le 3{,}1768 \tag{140}$$

where

$$\varphi_0(\mu_1) = \frac{\mu_1 - 1{,}5}{9{,}75 + 25{,}875\mu_1} \tag{140a}$$

$$\varphi_1(\mu_1) = \frac{9{,}75 + 25{,}875\mu_1}{9{,}75 + 19{,}375\mu_1 - 17{,}25\mu_1^2}\sqrt{\begin{array}{l}38{,}8125 - 51{,}75\mu_1 + \\ +17{,}25\mu_1^2\end{array}} \qquad (140b)$$

$$cos\alpha_{1,2} = 0{,}7494 \qquad (141)$$

$$\omega_2(\lambda_2, \alpha_{1,2}) = 0{,}0901 \cdot \varphi_1(\mu_1) \qquad (142)$$

Substituting the numerical expressions of A, λ_2, $cos\alpha_{1,2}$ and $\omega_2(\lambda_2, \alpha_{1,2,})$ Eqs. (139)–(142) in Eq. (134), find the final form of the vector function of the second piecewise-linear straight line depending on the first piecewise-linear straight line $\vec{z}_1$, and $cos\alpha_{1,2}$ in the form:

$$\vec{z}_2 = \vec{z}_1 \cdot \varphi_2(\mu_1) \text{ for } 1{,}5 \le \mu_1 \le 3{,}1768 \qquad (143)$$

where

$$\varphi_2(\mu_1) = 1 - 25{,}8751 \cdot \varphi_0(\mu_1) \cdot [1 + 0{,}0901 \cdot \varphi_1(\mu_1)] \qquad (144)$$

Note that the obtained Eq. (143) is a vector equation of the second straight line where the value of the parameter $\mu_1 \ge 1{,}5$. When we impose on the parameter μ_1 the condition $1{,}5 \le \mu_1 \le 3{,}1768$, Eq. (143) will represent a vector equation of the second piecewise-linear segment.

Calculate the value of the intersection point of the second and third piecewise-linear straight lines, i.e., at the point κ_2. Therefore, according to approximation of piecewise-linear straight lines, for the intersection point accept the value of the parameter $\mu_2^{\kappa_2} = 2$, and the approximate value of the parameter $\mu_1^{\kappa_2}$ calculated earlier will correspond to the upper value of inequality Eq. (133), i.e., $\mu_1^{\kappa_2} = 3{,}1768$. In this case, we find the value of the vector function $z_2^{k_2}$ in the coordinate form from Eq. (143) in the following form (Fig. 7):

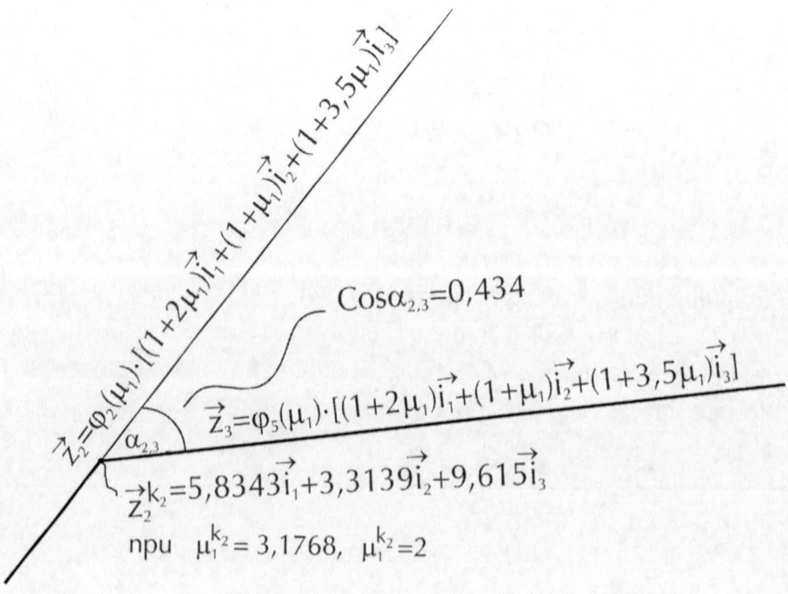

FIGURE 7 Numerical construction of 3-component piecewise-linear economic-mathematical model in 3-dimensional vector space R_3.

$$\vec{z}_2^{k_2} = 5{,}8343\vec{i}_1 + 3{,}3139\vec{i}_2 + 9{,}615\vec{i}_3 \tag{145}$$

Now construct a vector equation for the third piecewise-linear straight line depending on the vector equation of the first piecewise-linear function $\vec{z}_1$, $cos\alpha_{1,2}$, $cos\alpha_{2,3}$ and also parameter μ_1 corresponding to the parameter μ_3. For that we use the following defining Eqs. (87), (99)–(102):

$$\vec{z}_3 = \vec{z}_1\{1 + A[1 + \omega_2(\lambda_2^{k_2}, \alpha_{1,2}) + \omega_3(\lambda_3, \alpha_{2,3})]\}$$

for

$$\mu_1 \geq \mu_1^{k_2}, \ \mu_2 \geq \mu_2^{k_2} \tag{146}$$

Here the unaccounted parameters influence function $\omega_2(\lambda_2^{k_2},\alpha_{1,2})$ is calculated by means of Eq. (97), the unaccounted parameter influence function $\omega_3(\lambda_3,\alpha_{2,3})$ is calculated from Eqs. (100)–(102) in the form:

$$\omega_3(\lambda_3,\alpha_{2,3}) = \lambda_3 \cdot \cos\alpha_{2,3} \tag{147}$$

$$\lambda_3 = \frac{\mu_3}{\mu_1^{k_1} - \mu_1} \frac{\left|\vec{z}_2 - \vec{z}_2^{k_2}\right|\left|\vec{a}_4 - \vec{z}_2^{k_2}\right|}{\vec{z}_1(\vec{z}_2 - \vec{z}_2^{k_2})} \frac{\vec{z}_1(\vec{z}_1^{k_1} - \vec{a}_1)}{\left|\vec{a}_2 - \vec{a}_1\right|\left|\vec{z}_1^{k_1} - \vec{a}_1\right|} \tag{148}$$

$$\cos\alpha_{2,3} = \frac{(\vec{z}_2(\mu_2) - \vec{z}_2^{k_2})(\vec{a}_4 - \vec{z}_2^{k_2})}{\left|\vec{z}_2(\mu_2) - \vec{z}_2^{k_2}\right|\left|\vec{a}_4 - \vec{z}_2^{k_2}\right|}, \text{ for } \mu_2 \geq \mu_2^{k_2} \tag{149}$$

$$\mu_3 = (\mu_2 - \mu_2^{k_2})\frac{(\vec{a}_3 - \vec{z}_1^{k_1})(\vec{a}_4 - \vec{z}_2^{k_2})}{(\vec{a}_4 - \vec{z}_2^{k_2})^2}, \text{ for } \mu_2 \geq \mu_2^{k_2} \tag{150}$$

$$A = (\mu_1^{k_1} - \mu_1)\frac{\left|\vec{a}_2 - \vec{a}_1\right|\left|\vec{z}_1^{k_1} - \vec{a}_1\right|}{\vec{z}_1(\left|\vec{z}_1^{k_1} - \vec{a}_1\right|)} \tag{151}$$

$$\omega_2(\lambda_2^{k_2},\alpha_{1,2}) = \lambda_2^{k_2} \cdot \cos\alpha_{1,2} \tag{152}$$

Note that calculation of the function $\omega_2(\lambda_2^{k_2},\alpha_{1,2})$ is simplified owing to expression Eq. (142), where instead of the parameter μ_1 we should use its value corresponding to the second intersection point, i.e., $\mu_1 = \mu_1^{k_2} = 3{,}1768$. In this case, we get:

$$\omega_2(\lambda_2^{k_2},\alpha_{1,2}) = \omega_2(\lambda_2,\alpha_{1,2})\Big|_{\mu_1=3{,}1768} =$$

$$= 0{,}0901 \cdot \varphi_1(\mu_1)\Big|_{\mu_1=3{,}1768} = -0{,}5613 \tag{153}$$

The mathematical Eq. of the relation of the parameter μ_3 with the parameter μ_1 corresponding to the points of the first piecewise-linear straight line, will look like as follows. We have the condition of relation of the parameter μ_2 and the parameter μ_1 in the form Eq. (130). Therefore, substituting Eqs. (130) in Eq. (150), we get:

$$\mu_3 = [1{,}1927(\mu_1 - 1{,}5) - \mu_2^{k_2}]\frac{(\vec{a}_3 - \vec{z}_1^{k_1})(\vec{a}_4 - \vec{z}_2^{k_2})}{(\vec{a}_4 - \vec{z}_2^{k_2})^2} \tag{154}$$

Taking into account Eqs. (127), (129), (132), (145) for $\mu_2^{k_2} = 2$, Eq. (154) accepts the form:

$$\mu_3 = 0{,}4216 \cdot \mu_1 - 1{,}3394 \text{ for } \mu_3 \geq 0,\ \mu_1 \geq \mu_1^{k_2} = 3{,}1768 \tag{155}$$

Thus, Eq. (155) establishes the numerical relation between the parameters μ_3 and μ_1.

For $\mu_3 = 0$, $\mu_1|_{\mu_3=0} = \mu_1^{k_2} = 3{,}1768$, i.e., coincides with the value of the parameter $\mu_1^{k_2}$. For any values of the parameter $\mu_3 \geq 0$, the appropriate value of the parameter μ_1 will be greater than 3,1768, i.e., $\mu_1 \geq 3{,}1768$.

Substituting Eqs. (127), (129), (144), (145), (155) in Eq. (148), we get the numerical dependence of the unaccounted factors parameter on the third vector straight line λ_3 depending on the parameter μ_1 for $\mu_1 \geq \mu_1^{k_2} = 3{,}1768$ in the form:

$$\lambda_3 = -0{,}2356 \cdot \varphi_3(\mu_1), \text{ for } \mu_1 \geq \mu_1^{k_2} = 3{,}1768 \tag{156}$$

where the expression $\varphi_3(\mu_1)$ is of the form:

$$\varphi_3(\mu_1) = \frac{0{,}4216 \cdot \mu_1 - 1{,}3394}{\varphi_0(\mu_1)} \cdot$$

$$\cdot \frac{\sqrt{\varphi_2^2(\mu_1) \cdot [17{,}25\mu_1^2 + 13\mu_1 + 3] - 2\varphi_2(\mu_1)[48{,}635\mu_1 + 18{,}785] + 137{,}4692}}{\varphi_2(\mu_1) \cdot (17{,}25 \cdot \mu_1^2 + 13\mu_1 + 3) - 48{,}635 \cdot \mu_1 - 18{,}763} \tag{157}$$

Substituting Eqs. (127), (144), (145) in Eq. (149), we get the numerical dependence of $cos\alpha_{2,3}$ generated between the second and third piece-wise-linear straight lines depending on the parameter μ_1 for the values $\mu_1 \geq \mu_1^{k_2} = 3{,}1768$ in the form:

$$cos\alpha_{2,3} = 0{,}1640 \cdot \varphi_4(\mu_1), \text{ for } \mu_1 \geq 3{,}1768 \tag{158}$$

where the expression $\varphi_4(\mu_1)$ is of the form:

$$\varphi_4(\mu_1) = \frac{\begin{gathered}2{,}1657[\varphi_2(\mu_1)(1+2\mu_1)-5{,}8343]+\\+5{,}6861[\varphi_2(\mu_1)(1+\mu_1)-3{,}3139]+\\+0{,}385[\varphi_2(1+3{,}5\mu_1)-9{,}615]\end{gathered}}{\sqrt{\begin{gathered}[\varphi_2(\mu_1)(1+2\mu_1)-5{,}8343]^2+\\+[\varphi_2(\mu_1)(1+\mu_1)-3{,}3139]^2+\\[+\varphi_2(1+3{,}5\mu_1)-9{,}615]^2\end{gathered}}} \tag{159}$$

As the angle between the two straight lines is a constant quantity, we calculate the numerical value of $cos\alpha_{2,3}$ for $\mu_1 = 5$. In this case, we have (Fig. 7):

$$cos\alpha_{2,3} = 0{,}434 \tag{160}$$

Substituting Eqs. (156) and (160) in Eq. (147), the numerical value of the parameter $\omega_3(\lambda_3, \alpha_{2,3})$ will equal:

$$\omega_3(\lambda_3, \alpha_{2,3}) = \text{-}0{,}1022 \cdot \varphi_3(\mu_1) \text{ for } \mu_1 \geq 3{,}1768 \tag{161}$$

Now calculate the coefficient A Eq. (151) for $\mu_1 \geq \mu_1^{k_2} = 3{,}1768$. For that we substitute Eqs. (153), (161), and (162) in Eq. (151) and get the following numerical expression of the coefficient A:

$$A = -25{,}875 \frac{\mu_1 - 1{,}5}{9{,}75 + 25{,}875 \cdot \mu_1} = -25{,}875 \cdot \varphi_0(\mu_1),$$

for

$$\mu_1 \geq \mu_1^{k_2} = 3{,}1768 \tag{162}$$

or

$$A = -25{,}875 \cdot \varphi_0(\mu_1) \tag{163}$$

Substituting the numerical values Eqs. (153), (161) and (162) in Eq. (146), we get a vector equation for the third piecewise-linear straight line, expressed by the vector equation of the first piecewise-linear straight-line and the parameter μ_1 in the form (Fig. 7):

$$\vec{z}_3 = \vec{z}_1 \cdot \varphi_5(\mu_1), \text{ for } \mu_1 \geq \mu_1^{k_2} = 3{,}1768 \tag{164}$$

where

$$\varphi_5(\mu_1) = 1 - 11{,}3514 \cdot \varphi_0(\mu_1) \cdot [1 - 0{,}233 \cdot \varphi_3(\mu_1)] \tag{165}$$

or in the form:

$$\vec{z}_3 = \varphi_5(\mu_1) \cdot [(1 + 2\mu_1)\vec{i}_1 + (1 + \mu_1)\vec{i}_2 + (1 + 3{,}5\mu_1)\vec{i}_3] \tag{165.a}$$

Now investigate the prediction of economic process and its control on the subsequent $V_4(x_1, x_2, x_3)$ small volume of 3-dimensional vector space with regard to unaccounted parameter factors $\omega_2(\lambda_2^{k_2}, \alpha_{1,2})$ and $\omega_3(\lambda_3^{k_3}, \alpha_{2,3})$ that hold on the preceding stages of the process [4–6]. And the numerical values of these unaccounted parameters functions $\omega_2(\lambda_2, \alpha_{1,2})$, $\omega_2(\lambda_2^{k_2}, \alpha_{1,2})$ and $\omega_3(\lambda_3, \alpha_{2,3})$ are assumed to be known and are given by Eqs. (153), (161), and (142), having the following numerical expressions:

$$\omega_2(\lambda_2, \alpha_{1,2}) = 0{,}0901 \cdot \varphi_1(\mu_1) \tag{166}$$

$$\omega_2(\lambda_2^{k_2}, \alpha_{1,2}) = -0{,}5613 \tag{167}$$

$$\omega_3(\lambda_3,\alpha_{2,3}) = \text{-}0{,}1022 \cdot \varphi_3(\mu_1) \text{ for } \mu_1 \geq 3{,}1768 \tag{168}$$

where the expressions $\varphi_1(\mu_1)$ and $\varphi_3(\mu_1)$ are represented by Eqs. (140b) and (157).

Above for the 3-component piecewise-linear economic process we have constructed the third piecewise-linear straight line Eq. (164) depending on an arbitrary parameter μ_1 and unaccounted parameters influence spatial functions $\omega_2(\lambda_2,\alpha_{1,2})$ and $\omega_3(\lambda_3,\alpha_{2,3})$. On the other hand, by Eq. (115) we suggested for the 3-component case the economic process predicting vector function $\vec{Z}_4(\beta)$ with regard to the introduced unaccounted parameters predicting influence function $\Omega_4(\lambda_4,\alpha_{3,4})$ [4–6]:

$$\vec{Z}_4(\beta) = \vec{z}_1\{\ + A[1+\omega_2(\lambda_2^{k_2},\alpha_{1,2}) + \omega_3(\lambda_3^{k_3},\alpha_{2,3}) + \\ + \Omega_4(\lambda_4,\alpha_{3,4})]\}$$

for

$$\beta = 1{,}2 \tag{169}$$

$$\Omega_4(\lambda_4,\alpha_{3,4}) = \lambda_4 \cdot cos\alpha_{3,4} \tag{170}$$

$$\lambda_4 = \frac{\mu_4}{\mu_1^{k_1}-\mu_1} \frac{\left|\vec{z}_3-\vec{z}_3^{k_3}\right|\left|\vec{a}_5(\beta)-\vec{z}_3^{k_3}\right|}{\vec{z}_1(\vec{z}_3-\vec{z}_3^{k_3})} \frac{\vec{z}_1(\vec{z}_1^{k_1}-\vec{a}_1)}{\left|\vec{a}_2-\vec{a}_1\right|\left|\vec{z}_1^{k_1}-\vec{a}_1\right|},$$

for $$\beta = 1{,}2\,,\ \mu_3 \geq \mu_3^{k_3}, \mu_4 \geq 0 \tag{171}$$

$$\mu_4 = (\mu_1-\mu_3^{k_3})\frac{(\vec{a}_4-\vec{z}_2^{k_2})(\vec{a}_5(\beta)-\vec{z}_3^{k_3})}{(\vec{a}_5(\beta)-\vec{z}_3^{k_3})^2},$$

for $$\beta = 1{,}2\,,\ \mu_3 \geq \mu_3^{k_3}, \mu_4 \geq 0 \tag{172}$$

Here, $\omega_2(\lambda_2^{k_2},\alpha_{1,2})$ has numerical expression Eq. (167), the function $\omega_3(\lambda_3^{k_3},\alpha_{2,3})$ for the final point of the third piecewise-linear straight line for $\mu_3=\mu_3^{k_3}$ and its appropriate values $\mu_1=\mu_1^{k_3}$ is calculated by means of Eq. (168). As the intersection points of the straight lines are given, accept the value of the intersection point k_3 between the third and fourth predicting straight lines in the form $\mu_3^{k_3}=3$. And define the appropriate value of the parameter $\mu_1^{k_3}$ from the Eq. connecting the parameters μ_1 and μ_3 in the form Eq. (155):

$$\mu_3=0{,}4216\cdot\mu_1-1{,}3394,\ \text{for}\ \mu_3\geq 0,\ \mu_1\geq\mu_1^{k_2}=3{,}1768 \qquad (173)$$

Hence

$$\mu_1=2{,}3719\cdot\mu_3+3{,}1768 \qquad (174)$$

Substituting the value $\mu_3=\mu_3^{k_3}=3$ in Eq. (174), define the numerical value of the parameter $\mu_1^{k_3}$ corresponding to the value of the parameter $\mu_3^{k_3}=3$ at the intersection point of the third piecewise-linear straight line with the predicting fourth straight line in the form:

$$\mu_1^{k_3}=10{,}2926 \qquad (175)$$

Substituting the numerical value of $\mu_1^{k_3}$ Eq. (175) in Eq. (168), define the numerical value of the unaccounted parameters function $\omega_3(\lambda_3^{k_3},\alpha_{2,3})$ at the intersection point between the third piecewise-linear straight line and predicting fourth vector straight line. For that as preliminarily, by Eqs. (140a, b), (144) calculate the functions $\varphi_0(\mu_1)$, $\varphi_1(\mu_1)$ and $\varphi_2(\mu_2)$ for $\mu_1=\mu_1^{k_3}=10{,}2926$, and get:

$$\varphi_0(\mu_1)\Big|_{\mu_1=10,2926}=0{,}03185 \qquad (176)$$

$$\varphi_1(\mu_1)\Big|_{\mu_1=10,2926}=-6{,}23 \qquad (177)$$

$$\varphi_2(\mu_1)\Big|_{\mu_1=10,2926}=0{,}639 \qquad (178)$$

Now, substituting the numerical values Eqs. (176)–(178) in Eq. (168), take into account $\mu_1 = 10{,}2926$ and define the numerical value of the unaccounted parameter function $\omega_3(\lambda_3^{k_3}, \alpha_{2,3})$ at the third intersection point k_3 in the form:

$$\omega_3(\lambda_3^{k_3}, \alpha_{2,3}) = -0{,}2172 \tag{179}$$

Substituting Eq. (176)–(178) in Eq. (157) allowing for $\mu_1 = 10{,}2926$ we define the function $\varphi_3(\mu_1)\big|_{\mu_1=10,2926}$ in the form:

$$\varphi_3(\mu_1)\big|_{\mu_1=10,2926} = 2{,}125 \tag{180}$$

Substituting Eqs. (176) and (180) in Eq. (165), where we accept $\mu_1 = 10{,}2926$, find the numerical value of $\varphi_5(\mu_1)\big|_{\mu_1=10,2926}$ in the form:

$$\varphi_5(\mu_1)\big|_{\mu_1=10,2926} = 0{,}8175 \tag{181}$$

Substituting Eqs. (181) and (128) in Eq. (164) or Eq. (165a), where it is accepted $\mu_1 = 10{,}2926$, find the coordinate expression of the vector point $\vec{z}_3^{k_3}$ in the form (Fig. 8):

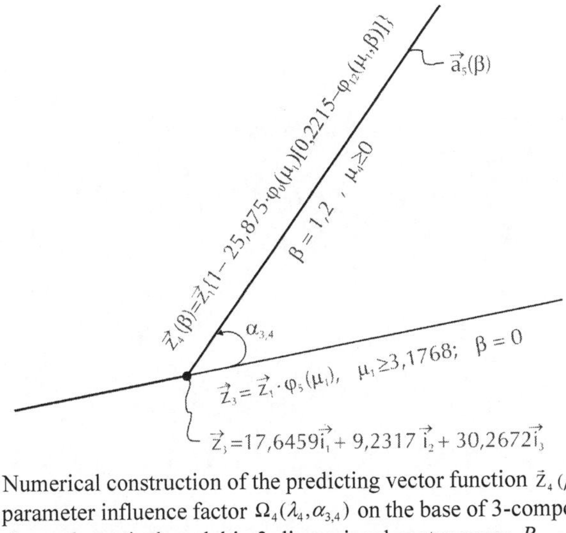

FIGURE 8 Numerical construction of the predicting vector function $\vec{z}_4(\beta)$ with regard to unaccounted parameter influence factor $\Omega_4(\lambda_4, \alpha_{3,4})$ on the base of 3-component piecewise-linear economic-mathematical model in 3-dimensional vector space R_3.

$$\vec{z}_3^{k_3} = 17{,}6459 \cdot \vec{i}_1 + 9{,}2317 \cdot \vec{i}_2 + 30{,}2672 \cdot \vec{i}_3 \tag{182}$$

Now, by Eq. (170) calculate the unaccounted parameters predicting function $\Omega_4(\lambda_4, \alpha_{3,4})$. For defining it, as preliminarily we find the numerical dependence of the parameter μ_4 on the parameter μ_1, λ_4, $cos\alpha_{3,4}$, and also on the vector $\vec{a}_5(\beta)$ for $\beta = 1{,}2$. Therewith we note that the vector $\vec{a}_5(\beta)$ for the values $\beta = 1{,}2$ has coordinate form Eq. (123). Here the coordinate $\vec{a}_{5i}(\beta)$ in 3-dimensional space are determined by Eq. (124). Substituting Eqs. (124), (127), (145), (173) in Eq. (172) for $\mu_3^{k_3} = 3$ establish the numerical dependence of the parameter μ_4 on the parameter μ_1 in the form:

$$\mu_4 = (0{,}4216 \cdot \mu_1 - 4{,}3393) \cdot \varphi_6(\mu_1, \beta)$$
$$\text{for } \mu_1 \geq 10{,}2926, \; \beta = 1{,}2 \tag{183}$$

where

$$\varphi_6(\mu_1, \beta) =$$
$$= \frac{2{,}1657[a_{51}(\beta) - \varphi_5(1+2\mu_1)] + 5{,}6861[a_{52}(\beta) - \varphi_5(1+\mu_1)] + 0{,}385[a_{53}(\beta) - \varphi_5(1+3{,}5\mu_1)]}{[a_{51}(\beta) - \varphi_5(1+2\mu_1)]^2 + [a_{52}(\beta) - \varphi_5(1+\mu_1)]^2 + [a_{53}(\beta) - \varphi_5(1+3{,}5\mu_1)]^2} \tag{184}$$

Substitute Eqs. (122), (127)–(129), (164), (182), and (183) allowing for Eq. (124) in Eq. (171) and define the predicting parameter λ_4 in the form:

$$\lambda_4 = -\varphi_8(\mu_1) \cdot \varphi_6(\mu_1, \beta) \cdot \varphi_7(\mu_1, \beta) \tag{185}$$

where

$$\varphi_7(\mu_1, \beta) = \sqrt{[a_{51}(\beta) - 17{,}6459]^2 + [a_{51}(\beta) - 9{,}2317]^2 + [a_{53}(\beta) - 30{,}2672]^2} \tag{186}$$

$$\varphi_8(\mu_1) = \frac{\sqrt{\begin{array}{l}[(1+2\mu_1)\varphi_5 - 17{,}6459]^2 + \\ +[(1+\mu_1)\varphi_5 - 9{,}2317]^2 + \\ +[(1+3{,}5\mu_1)\varphi_5 - 30{,}2672]^2\end{array}}}{\begin{array}{l}\varphi_5[(1+2\mu_1)^2 + (1+\mu_1)^2 + (1+3{,}5\mu_1)^2] - \\ -[17{,}6459(1+2\mu_1) + \\ +9{,}2317(1+\mu_1) + 30{,}2672(1+3{,}5\mu_1)]\end{array}} \tag{187}$$

Now define the numerical value of $cos\alpha_{3,4}$ generated between the third piecewise-linear function $\vec{z}_3$ and predicting fourth vector function $\vec{Z}_4(\mu_4)$ (Fig. 8) [18–21].

$$cos\alpha_{3,4} = \frac{(\vec{z}_3 - \vec{z}_3^{k_3})(\vec{a}_5(\beta) - \vec{z}_3^{k_3})}{\left|\vec{z}_3 - \vec{z}_3^{k_3}\right|\left|\vec{a}_5(\beta) - \vec{z}_3^{k_3}\right|} \tag{188}$$

For that substitute Eqs. (123), (128), (164), and (182) in Eq. (188) and get:

$$cos\alpha_{3,4} = \frac{\varphi_9(\mu_1, \beta)}{\varphi_{10}(\mu_1, \beta)\cdot\varphi_{11}(\mu_1, \beta)} \tag{189}$$

where

$$\begin{array}{l}\varphi_9(\mu_1, \beta) = \mathrm{a}_{51}(\beta)[(1+2\mu_1)\varphi_5 - 17{,}6459] + \\ +\mathrm{a}_{52}(\beta)[(1+\mu_1)\varphi_5 - 9{,}2317] + \\ +\mathrm{a}_{53}(\beta)[(1+3{,}5\mu_1)\varphi_5 - 30{,}2672] - \\ -\varphi_5[17{,}6459(1+2\mu_1) + 9{,}2317(1+\mu_1) + \\ +30{,}2672(1+3{,}5\mu_1)] + 1312{,}7085\end{array} \tag{190}$$

$$\varphi_{10}(\mu_1,\beta)=$$

$$=\sqrt{\begin{aligned}&\varphi_5^2[(1+2\mu_1)^2+(1+\mu_1)^2+(1+3{,}5\mu_1)^2]-\\&-2\varphi_5[17{,}6459(1+2\mu_1)+\\&+9{,}2317(1+\mu_1)+30{,}2672(1+3{,}5\mu_1)]+1312{,}7055\end{aligned}} \tag{191}$$

$$\varphi_{11}(\mu_1,\beta)=\sqrt{\begin{aligned}&[a_{51}(\beta)-17{,}6459]^2+[a_{51}(\beta)-9{,}2317]^2+\\&+[a_{53}(\beta)-30{,}2672]^2\end{aligned}} \tag{192}$$

Substituting Eqs. (185) and (189) in Eq. (170), establish the numerical representation of the unaccounted parameter predicting influence function $\Omega_4(\lambda_4,\alpha_{3,4})$ in the form:

$$\Omega_4(\lambda_4,\alpha_{3,4})=-\varphi_{12}(\mu_1,\beta) \tag{193}$$

where

$$\varphi_{12}(\mu_1,\beta)=$$

$$=\varphi_6(\mu_1,\beta)\cdot\varphi_7(\mu_1,\beta)\cdot\varphi_8(\mu_1)\cdot\frac{\varphi_9(\mu_1,\beta)}{\varphi_{10}(\mu_1,\beta)\cdot\varphi_{11}(\mu_1,\beta)} \tag{194}$$

Representing the numerical values Eqs. (153), (162), (179), and (193) in Eq. (169), define the concrete form of the predicting vector function on the fourth small volume of 3-dimensional space for $\beta=1{,}2$ in the form (Fig. 8):

$$\vec{Z}_4(\beta)=\vec{z}_1\{1-25{,}875\cdot\varphi_0(\mu_1)[1-0{,}5613-0{,}2172-\varphi_{12}(\mu_1,\beta)]\}$$

for $\beta=1{,}2$

or

$$\vec{Z}_4(\beta) = \vec{z}_1\{1 - 25{,}875 \cdot \varphi_0(\mu_1)[0{,}2215 - \varphi_{12}(\mu_1, \beta)]\}$$

for $\beta = 1{,}2$ (195)

or in the coordinate form:

$$\vec{Z}_4(\beta) = \{1 - 25{,}875 \cdot \varphi_0(\mu_1)[0{,}2215 - \\ - \varphi_{12}(\mu_1, \beta)]\} \cdot [(1 + 2\mu_1)\vec{i}_1 + (1 + \mu_1)\vec{i}_2 + (1 + 3{,}5\mu_1)\vec{i}_3] \tag{196}$$

for $\mu_1 \geq 10{,}2926$, $\beta = 1{,}2$

KEYWORDS

- **2-component piecewise-linear economic mathematical model**
- **3-component piecewise-linear economic mathematical model**
- **3-dimensional vector space**
- **piecewise-linear straight lines**

REFERENCES

1. Aliyev Azad, G. *Perspectives of development of the methods of modeling and prediction of economic processes.* Izv. NAS of Azerbaijan, Ser. of hummanitarian and social sciences (economics), Baku, **2006,** *2,* 253–260.
2. Aliyev Azad, G. *A construction methods for piecewise-linear economic-mathematical models with regard to uncertainty factor in finite-dimensional vector space.* Collection of papers of the Institute of Economics of NAS of Azerbaijan (Problems of national economics), ISБN-5-8066-1711-3, Baku, **2007,** *2,* 290–301.
3. Aliyev Azad, G. *Theoretical aspects of the problem of analysis of risks of investigation projects in oilgas recovery industry.* Izv. Of NASA.; Ser. of human and social sciences (economics), Baku, **2009,** 1, 97–104.
4. Aliyev Azad, G. *2-component piecewise-linear economic mathematical model and economic event prediction methods in uncertainty conditions in 3-dimensional vector*

space. "Problem Economical" *(2),* Izv."Sputnik" publishing. ISSN 1813–8578, Moscow, **2009,** 111–124.

5. Aliyev Azad, G. *Theoretical bases of economic mathematical simulation at uncertainty conditions*, National Academy of Sciences of Azerbaijan, Baku, "Information Technologies," ISBN 995221056-9, **2011,** 338.
6. Aliyev Azad, G. *Economical and mathematical models subject to complete information.* Ed. Lap Lambert Akademic Publishing. ISBN 978-3-659-30998-4, Berlin, **2013,** 316.
7. Aliyev Azad, G. *On a criterion of economic process certainty in finite-dimensional vector space.* "Economic, Statistic and Informatical." Vestnik of UMO; Moscow, **2008,** *2*, 33–37.
8. Aliyev Azad, G. *Some problems of prediction and control of economic event in uncertainty conditions in 3-dimensional vector space.* Proceedings of the II International scientific-practical Conference "Youth and science: reality and future," Natural and Applied Sciences. Nevinnomysk, **2009,** *8,* 394–396.
9. Aliyev Azad, G. *Bases of piecewise-linear economic-mathematical models with regard to unaccounted factors influence on a plane and multivariant prediction of economic event.* "Issues of economics science" "Sputnik," ISSN 1728–8878, Moscow, **2009,** 187–201.
10. Aliyev Azad, G. *On a principle of prediction and control of economic process with regard to uncertainty factor in 1-dimensional vector space,* "Economic, Statistical and Informatical" Vestnik, UMO: Moscow, **2008,** *4,* 27–32.
11. Aliyev Azad, G. *Economic-mathematical methods and models in uncertainty conditions in finite-dimensional vector space.* Ed. NAS of Azerbaijan "Information technologies." ISBN 978-9952-434-10-1, Baku, **2009,** 3, 220.
12. Aliyev Azad, G. *Construction of 2-component economic-mathematical models in complete information conditions in 3-dimensional vector space.* Izvestia of NAS of Azerbaijan, ser. of humanitarian and social sciences (economics). **2008,** *2*, Baku, 96–102.
13. Aliyev Azad, G. *Construction of 3-component economic-mathematical model in incomplete information conditions in 3-dimensional vector space.* Scientific proceeding of Nakhchivan State University. **2008,** *5 (25)*, 156–160.
14. Aliyev Azad, G. *Prediction of economic process in incomplete information conditions in 3-dimensional vector space on the base of 2-component economic-mathematical model.* "Igdisadiyyat ve Hayat," ISSN 0207–3021, Baku, **2008,** *9,* 35–40.
15. Aliyev Azad, G. *Prediction of economic process in incomplete information conditions in 3-dimensional vector space on the base of 3-component economic-mathematical model.* "Igdisadiyyat ve Hayat," ISSN 0207–3021, Baku, **2008,** *10,* 53–58.
16. Aliyev Azad, G. *Numerical calculation of 2-component piecewise-linear economic-mathematical model in 3-dimensional vector space.* Izvestia NAS of Azerbaijan, Baku, **2008,** *1,* 96–102.
17. Aliyev Azad, G. *Numerical calculation of economic event prediction with regard to uncertainty factor in 3-dimensional vector space on the base of 3-component piecewise-linear economic-mathematical model.* Collection of papers of the Institute of Economics NAS of Azerbaijan, Baku, **2008,** 129–143.

18. Aliyev Azad, G. *Difficulties of construction of economic-mathematical models, prediction and control of economic event in incomplete information conditions.* "Urgent problems of economics of transitional period" (collection of scientific papers). ASOA publishing, Baku, **2002,** 318–323.
19. Aliyev Azad, G. *On a mathematical aspect of constructing a conjugate vector for defining the unaccounted factors influence function of economic process in finite-dimensional vector space,* "Urgent social-economic problems of oil, chemical and machine-building fields of economics" (collection of scientific papers of the faculty "International Economic relations and Management" of Azerbaijan State Oil Academy) Baku, **2007,** 600–616.
20. Aliyev Azad, G. *Numerical calculation of economic event prediction with regard to uncertainty factor in 3-dimensional vector space on the base of 2-component piecewise-linear economic-mathematical model.* Collection of papers of the Institute of Economics of NAS of Azerbaijan (Problems of National Economics), ISBN-5-8066-1711-3, Baku, **2008,** *1,* 142–156.
21. Aliyev Azad, G. *Numerical calculation of 3-component piecewise-linear economic-mathematical model with regard to uncertainty factor in 3-dimensional vector space.* Izvestiya NAS of Azerbaijan, Ser. of human and social sciences (economics), Baku, **2008,** *3,* 86–100.

CHAPTER 4

PIECEWISE-LINEAR ECONOMIC-MATHEMATICAL MODELS WITH REGARD TO UNACCOUNTED FACTORS INFLUENCE ON A PLANE

CONTENTS

The theory developed in Chapter 2 will be applied to economic-mathematical problems on a plane [1–25].

In the coordinate variant there will be:

- given geometric interpretation of the introduced unaccounted parameter λ_m and unaccounted factors influence function $\omega_n(t,\lambda_n)$;
- suggested a simplified method of numerical calculation for economic problems on a plane with regard to unaccounted factors influence;
- approved a method for construction on a plane, piecewise-linear economic-mathematical aim functions on the model problems;
- suggested an appropriate method on multivariant prediction and control of economic event for plane problems.

4.1 A METHOD FOR CONSTRUCTING PIECEWISE-LINEAR ECONOMIC-MATHEMATICAL MODELS WITH REGARD TO UNACCOUNTED FACTORS INFLUENCE ON A PLANE: A WAY FOR DEFINING UNACCOUNTED FACTORS INFLUENCE FUNCTION

Given the statistical table (t_i,λ_i) describing some economic process. By the known appropriate methods approximate this system by n piecewise-homogeneous linear functions $y_n = y_n(t,\lambda_n)$.

Note that from the principle of certainty of economic process it follows that at the points of each piecewise interval (t_n, t_{n+k}), each of the considered functions $y_n = y_n(t,\lambda_n)$ are homogeneous in themselves, i.e., the function is analytic, the derivative of this function $y_n' = y_n'(t,\lambda_n) = \text{const}$. And the homogeneity degree of each function is different.

This means that each piecewise-linear function was obtained by observing definite external factors inherent to certain time interval. Thus, at different time intervals the economic process will have the influence of preceding external factors. Taking into account what has been said above, we give a method for constructing on a plane, a unique piecewise-linear function expressed by the first piecewise-linear function $y_1 = y_1(t)$ and arising additional external factors influence functions $\omega_n(t,\lambda_n)$. We call the functions $\omega_n(t,\lambda_n)$ as the unaccounted parameters functions or the external factors influence functions. Origination of the functions $\omega_n(t,\lambda_n)$ is connected with the fact that the rates of economic process at different time intervals are not the same [1–25].

So, represent any piecewise-homogeneous linear function $y_n = y_n(t, \lambda_n)$ by the first piecewise-homogeneous linear function $y_1 = y_1(t)$ and all $\omega_n(t_{n+1}, \lambda_n)$ unaccounted parameters influence functions influencing on the preceding general time interval (t_1, t_N).

According to Fig. 1, continue the first straight line to the point M_1 with appropriate coordinates $t_2 \leq t \leq t_3$. Draw a perpendicular M_1Q_1 on the axis t. Denote the ratio $\frac{M_1N_1}{M_1B_1}$ by $\omega_2(t, \lambda_2)$:

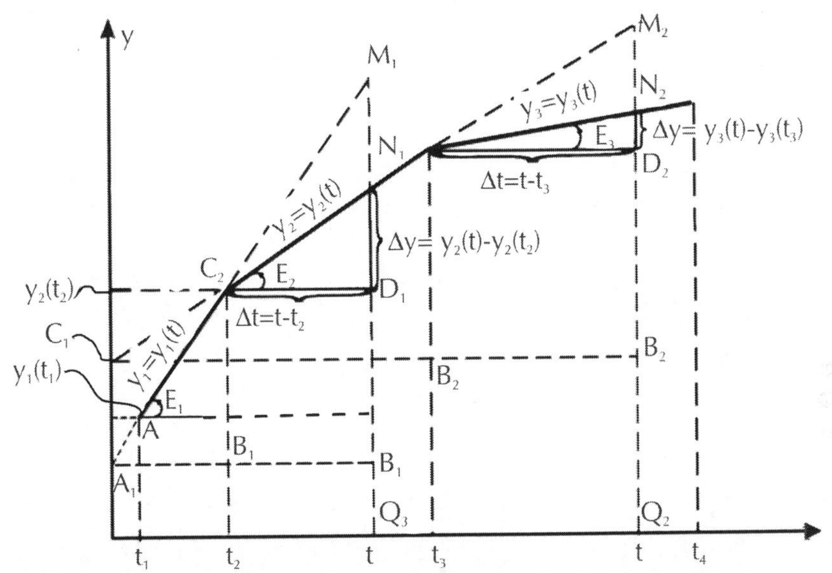

FIGURE 1 A scheme for constructing piecewise-linear economic-mathematical model on a plane.

$$\omega_2(t, \lambda_2) = \frac{M_1N_1}{M_1B_1} \tag{1}$$

In this case, for the points located in the interval (t_2 £ t £ t_3), we will have:

$$\frac{\Delta y}{\Delta t} = \frac{y_2(t) - y_2(t_2)}{t - t_2} = E_2 \tag{2}$$

$$y_1(t) = E_1(t - t_1) + y_1(t_1) \tag{3}$$

From Eqs. (2) and (3) allowing for the conjugation condition of piecewise-linear functions at the point $t = t_2$, i.e., $y_1(t_2) = y_2(t_2)$ we get:

$$y_2(t) = E_2(t - t_2) + E_1 t_2 - E_1 t_1 + y_1(t_1) \tag{4}$$

Now, in the right side of the obtained relation, add and subtract the expression $(\pm E_1 t)$, and get:

$$y_2(t) = E_1\left[1 - (1 - \frac{E_2}{E_1})(1 - \frac{t_2}{t})\right]t - E_1 t_1 + y_1(t_1) \tag{4a}$$

Here, E_1 and E_2 are tangents of the slopes of straight lines $y_1 = y_1(t)$ and $y_2 = y_2(t)$; $\lambda_2 = \frac{E_1 - E_2}{E_1}$ is the so called unaccounted factor influence parameter influencing on the second piecewise-linear function on time interval $(t_2 \le t \le t_3)$.

Introduce the denotation of the parameter of the form:

$$\omega_2(t, \lambda_2) = \frac{E_1 - E_2}{E_1}(1 - \frac{t_2}{t}) = \lambda_2(1 - \frac{t_2}{t}) \tag{5}$$

Determine the dependence of the function $y_2(t)$ for all the points of the second piecewise-linear straight line on difference factor of the economic process rate and the form of the first straight line in the following form:

$$y_2(t, \lambda_2) = E_1[1 - \omega_2(t, \lambda_2)]t + y_1(t_1) - E_1 t_1 \tag{6}$$

Here we call the function $\omega_2(t, \lambda_2)$ an unaccounted parameters influence function influencing on the point of the second piecewise-linear straight line on time interval $(t_2 \le t \le t_3)$.

Furthermore:

$$\omega_2(t, \lambda_2) = 0, \quad t \le t_2$$

$$\omega_2(t_3,\lambda_2)=\lambda_2(1-\frac{t_2}{t_3})=\frac{E_1-E_2}{E_1}(1-\frac{t_2}{t_3})\text{, for } t=t_3$$

Special case. If the first piecewise-linear function passes through the origin of coordinates, then $y_1(t_1)=0_1$ for $t_1=0$, and Eq. (4.1.4) accepts a simpler form:

$$y_2(t)=E_1[1-\omega_2(t,\lambda_2)]t \tag{6a}$$

4.2 GEOMETRICAL INTERPRETATION OF THE UNACCOUNTED FACTORS INFLUENCE FUNCTION ON A PLANE

The introduced unaccounted parameters influence function $\omega_2(t,\lambda_2)$ is of the form Eq. (5):

$$\omega_2(t,\lambda_2)=\lambda_2(1-\frac{t_2}{t})=(1-\frac{E_2}{E_1})(1-\frac{t_2}{t})$$

Prove that in the geometrical sense, Eq. (5) is the ratio of the segment M_1N_1 to M_1B_1 at each point of the second piecewise-linear straight line, i.e.,

$$\omega_2(t,\lambda_2)=\frac{M_1N_1}{M_1B_1} \tag{7}$$

From Fig. 1 of Chapter 5, we have:

$$\omega_2(t,\lambda_2)=\frac{M_1N_1}{M_1B_1}=\frac{M_1B_1-N_1D_1-D_1B_1}{M_1B_1} \tag{8}$$

From D $A_1M_1B_1$ we have:

$$A_1B_1=t\text{, }\frac{M_1B_1}{A_1B_1}=E_1\text{, }M_1B_1=E_1t$$

From DN_1D_1C we have:

$$CD_1 = \Delta t = t - t_2, \quad \frac{N_1D_1}{CD_1} = E_2,$$

$$N_1D_1 = E_2(t - t_2) = E_2\Delta t, \; D_1B_1 = CB_1' = E_1t_2$$

Taking into account the expressions of M_1B_1, N_1D_1 and D_1B_1, eq. (8) will take the form:

$$\omega_2(t, \lambda_2) = \frac{E_1t - E_2\Delta t - E_1t_2}{E_1t} = \frac{E_1 - E_2}{E_1} \cdot \frac{\Delta t}{t} = \frac{E_1 - E_2}{E_1}(1 - \frac{t_2}{t})$$

that coincides with the theoretically obtained eq. (5) Q.E.D.

Continue the second piecewise-homogeneous linear function to the point M_2 corresponding to the coordinate t ($t_3 \le t \le t_4$). Draw the perpendicular M_2Q_2 to the axis t. Denote the ratio $\frac{M_2N_2}{M_2B_2}$ by $\omega_3(t, \lambda_3)$:

$$\omega_3(t, \lambda_3) = \frac{M_2N_2}{M_2B_2} \tag{9}$$

For the points t situated in the time interval t_3 £ t £ t_4, from (Fig. 1) we have:

$$\frac{\Delta y}{\Delta t} = \frac{y_3(t) - y_3(t_3)}{t - t_3} = E_3 \tag{10}$$

$$y_2(t, \lambda_2) = E_1\,[1 - \omega_2(t, \lambda_2)]\, t + y_1(t_1) - E_1t_1. \tag{11}$$

From Eqs. (10) and (11) allowing for the conjugation condition of piecewise-homogeneous linear functions at the point $t = t_3$, $y_2(t_3) = y_3(t_3)$, we get:

$$y_3(t) = E_3(t - t_3) + E_1[1 - \omega_2(t_3, \lambda_2)]t_3 + y_1(t_1) - E_1t_1$$

Now to the right side of the obtained relation add and subtract the expression $\pm E_1[1 - \omega_2(t_3, \lambda_2)]t$, and get:

$$y_3(t) = E_1 t\left\{[1-\omega_2(t_3,\lambda_2)]\left[1-\frac{1-\omega_2(t_3,\lambda_2)-\frac{E_3}{E_1}}{1-\omega_2(t_3,\lambda_2)}(1-\frac{t_3}{t})\right]\right\}+$$

$$+y_1(t_1)-E_1t_1 \tag{12}$$

Here E_3 is the tangent of the slope of the third piecewise-homogeneous linear function $y_3 = y_3(t)$;

$$\lambda_3 = \frac{1-\omega_2(t_3,\lambda_2)-\frac{E_3}{E_1}}{1-\omega_2(t_3,\lambda_2)} \tag{13}$$

Here λ_3 is the so-called dimensionless parameter of unaccounted factors influence parameter influencing on the third piecewise-homogeneous linear function on time interval $t_3 \le t \le t_4$;

$$\omega_3(t,\lambda_3) = \lambda_3(1-\frac{t_3}{t}) = \frac{1-\omega_2(t_3,\lambda_2)-\frac{E_3}{E_1}}{1-\omega_2(t_3,\lambda_2)}(1-\frac{t_3}{t}) \tag{14}$$

Here $\omega_3(t,\lambda_3)$ is the unaccounted parameter influence function influencing on the third piecewise-homogeneous linear function on time interval $t_3 \le t \le t_4$. furthermore:

$$\omega_3(t,\lambda_3) = 0\,,\ t \le t_3$$

$$\omega_3(t_4,\lambda_3) = \lambda_3(1-\frac{t_3}{t_4}) = \frac{1-\omega_2(t_3,\lambda_2)-\frac{E_3}{E_1}}{1-\omega_2(t_3,\lambda_2)}(1-\frac{t_3}{t_4}),\ \text{for } t = t_4$$

Thus, taking into account the introduced unaccounted parameter function $\omega_3(t,\ \lambda_3)$, we define the dependence of the points of the third piecewise-linear straight line on the rate difference factor of economic process between the second and third straight lines, and also on the form of the first straight line, in the form:

$$y_3(t,\lambda_3)=E_1[1-\omega_2(t_3,\lambda_2)]\cdot[1-\omega_3(t\ ,\lambda_3)]t+y_1(t_1)\quad E_1t_1 \tag{15}$$

Carrying out the similar operation for the fourth piecewise-homogeneous straight line, the dependence for the points of the fourth straight line on the rate difference factor of economic process between the third and fourth straight lines, and also on the form of the first straight line, will be in the form:

$$\begin{gathered}y_4(t,\lambda_4)=E_1[1-\omega_2(t_3,\lambda_2)]\cdot[1-\omega_3(t_4,\lambda_3)][1-\omega_4(t\ ,\lambda_4)]t+\\+y_1(t_1)-E_1t_1\end{gathered} \tag{16}$$

Here E_4 is the tangent of the slope of the fourth piecewise-homogeneous linear function $y_4=y_4(t)$;

$$\lambda_4=\frac{[1-\omega_2(t_3,\lambda_2)][1-\omega_3(t_4,\lambda_3)]-\dfrac{E_4}{E_1}}{[1-\omega_2(t_3,\lambda_2)][1-\omega_3(t_4,\lambda_3)]} \tag{17}$$

Here λ_4 is the influence parameter of unaccounted factors influencing on the fourth piecewise-homogeneous linear function on time interval $t_4\le t\le t_5$, and

$$\begin{gathered}\omega_4(t,\lambda_4)=\lambda_4(1-\frac{t_4}{t})=\\=\frac{[1-\omega_2(t_3,\lambda_2)][1-\omega_3(t_4,\lambda_3)]-\dfrac{E_4}{E_1}}{[1-\omega_2(t_3,\lambda_2)][1-\omega_3(t_4,\lambda_3)]}\cdot(1-\frac{t_4}{t})\end{gathered} \tag{18}$$

Here $\omega_4(t,\lambda_4)$ is the unaccounted parameter influence function influencing on the fourth piecewise-linear function on time interval $t_4 \le t \le t_5$.

Furthermore:

$$\omega_4(t,\lambda_3)=0,\ t\le t_4$$

$$\omega_4(t_5,\lambda_4)=\lambda_4(1-\frac{t_4}{t_5})=$$

$$=\frac{[1-\omega_2(t_3,\lambda_2)][1-\omega_3(t_4,\lambda_3)]-\frac{E_4}{E_1}}{[1-\omega_2(t_3,\lambda_2)][1-\omega_3(t_4,\lambda_3)]}\cdot(1-\frac{t_4}{t_5})$$

By recurrent way we can easily get the dependence of any piecewise-homogeneous linear function $y_n(t_n,\lambda_n)$ on the first piecewise-homogeneous linear function and all unaccounted parameters influence functions $\omega_n(t_{n+1},\lambda_n)$, influencing on the preceding common time interval (t_1,t_n) in the form:

$$y_n(t,\lambda_n)=E_1t[1-\omega_n(t,\lambda_n)]\prod_{k=2}^{n-1\ge 2}[1-\omega_k(t_{k+1},\lambda_k)]+ \tag{19}$$

$$+y_1(t_1)-E_1t_1,$$

where:

$$\lambda_n=\frac{\prod_{k=2}^{n-1\ge 2}[1-\omega_k(t_{k+1},\lambda_k)]-E_n/E_1}{\prod_{k=2}^{n-1\ge 2}[1-\omega_k(t_{k+1},\lambda_k)]} \tag{20}$$

$$\omega_n(t,\lambda_n)=\lambda_n(1-\frac{t_n}{t})=\frac{\prod_{k=2}^{n-1\geq 2}[1-\omega_k(t_{k+1},\lambda_k)]-E_n/E_1}{\prod_{k=2}^{n-1\geq 2}[1-\omega_k(t_{k+1},\lambda_k)]}(1-\frac{t_n}{t}) \tag{21}$$

Here:

$$\omega_n(t,\lambda_n)=0\text{, where } t\leq t_n$$

$$\omega_n(t_{n+1},\lambda_n)=\frac{\prod_{k=2}^{n-1\geq 2}[1-\omega_k(t_{k+1},\lambda_k)]-E_n/E_1}{\prod_{k=2}^{n-1\geq 2}[1-\omega_k(t_{k+1},\lambda_k)]}(1-\frac{t_n}{t_{n+1}})\text{, where } t=t_n$$

4.3 PREDICTION OF ECONOMIC EVENT AND ITS CONTROL BY PIECEWISE-LINEAR FUNCTIONS WITH REGARD TO UNACCOUNTED FACTORS INFLUENCE ON A PLANE

The problem on prediction of economic event and its control on the subsequent time interval (t_{N+1},t_{N+2}) by means of unaccounted factors influence function $\omega_m(t_{m+1},\lambda_m)$ on the preceding time interval (t_1,t_N) is solved as follows [1–3, 6].

Here we should note that the unaccounted parameters influence functions $\omega_m(t,\lambda_m)$ are integral characteristics of influencing external factors that in a priori not being in the structured model appear under the events occurring in external medium and render very strong functional influence both on the form of the function and the results of prediction quantities. And this cause ability can't be fixed by statistical means. In this connection, on the subsequent time interval (t_{N+1},t_{N+2}) of economic process, very likely we can see any of enumerated invisible factors or their combinations earlier hold in preceding time interval (t_1,t_N). Therefore, studying the prediction problem of any economic process on the subsequent time interval (t_{N+1}, t_{N+2}), it is necessary to be ready to possible influence of such factors. In connection with such a statement of the problem, let's investigate behavior of the economic process on the subsequent time interval situated under the desired unaccounted parameters influence function that we met at the preceding common time interval (t_1, t_N). For that construct the mathematical form of the aim function $y_{N+1}(t, \lambda_{N+1})$ depending on the first piecewise-linear function $y_1 = y_1(t)$ and the desired influence function $\omega_m(t, \lambda_m)$ that we met at the preceding common time interval (t_1, t_N). For that in Eqs. (19)–(21) replace $n = N$ by $n = (N+1)$, and get:

$$y_{N+1}(t,\lambda_{N+1}) = E_1 t[1-\omega_{N+1}(t,\lambda_{N+1})]\prod_{k=2}^{N}[1-\omega_k(t_{k+1},\lambda_k)] + \\ + y_1(t_1) - E_1 t_1 \tag{22}$$

where

$$\lambda_{N+1} = \frac{\prod_{k=2}^{N}[1-\omega_k(t_{k+1},\lambda_k)] - \dfrac{E_{N+1}}{E_1}}{\prod_{k=2}^{N}[1-\omega_k(t_{k+1},\lambda_k)]} \tag{23}$$

$$\omega_{N+1}(t,\lambda_{N+1}) = \lambda_{N+1}\left(1-\frac{t_{N+1}}{t}\right) = \\ = \frac{\prod_{k=2}^{N}[1-\omega_k(t_{k+1},\lambda_k)] - \dfrac{E_{N+1}}{E_1}}{\prod_{k=2}^{N}[1-\omega_k(t_{k+1},\lambda_k)]}\left(1-\frac{t_{N+1}}{t}\right) \tag{24}$$

A necessary influence condition of unaccounted parameter influence function $\omega_m(t, \lambda_m)$ is suggested in the form:

$$\omega_{N+1}(t,\lambda_{N+1})=\omega_m(\xi,\lambda_m) \tag{25}$$

or

$$\lambda_{N+1}(1-\frac{t_{N+1}}{t})=\lambda_m(1-\frac{\xi_m}{\xi}) \tag{26}$$

Here, the variable x changes in time interval $x_m \le x \le x_{m+1}$ corresponding to the action time of the influence function $\omega_m(t,\lambda_m)$.

As the influence is planned on time interval (t_{N+1},t_{N+2}), beginning with the point t_{N+1}, and also taking into account that in Eq. (26) the left side depends only on t, and the right side only on x, and also taking into account that for $t=t_{N+1}$ we have the condition $\omega_{N+1}(t,\lambda_{N+1})=0$, the necessary condition for the existence of equality Eq. (26) will be:

$$t=\xi,\ t_{N+1}=\xi_m,\ \lambda_{N+1}=\lambda_m.$$

Then, we will write Eqs. (22)–(24) in the form:

$$y_{N+1}(t,\lambda_{N+1})=E_1t[1-\omega_m(t,\lambda_m)]\prod_{k=2}^{N}[1-\omega_k(t_{k+1},\lambda_k)]+ \\ +y_1(t_1)-E_1t_1 \tag{27}$$

$$\lambda_m=\frac{\prod\limits_{\alpha=2}^{m-1\ge 2}[1-\omega_\alpha(t_{\alpha+1},\lambda_\alpha)]-\frac{E_m}{E_1}}{\prod\limits_{\alpha=2}^{m-1\ge 2}[1-\omega_\alpha(t_{\alpha+1},\lambda_\alpha)]} \tag{28}$$

$$\omega_m(t,\lambda_m)=\lambda_m(1-\frac{t_{N+1}}{t})= \\ =\frac{\prod\limits_{\alpha=2}^{m-1\ge 2}[1-\omega_\alpha(t_{\alpha+1},\lambda_\alpha)]-\frac{E_m}{E_1}}{\prod\limits_{\alpha=2}^{m-1\ge 2}[1-\omega_\alpha(t_{\alpha+1},\lambda_\alpha)]}(1-\frac{t_{N+1}}{t}) \tag{29}$$

The case $\omega_m(t,\lambda_m)=0$ for $t=t_{N+1}$ corresponds to the case when the external influences (external unaccounted factors) on time interval (t_{N+1},t_{N+2}) are the same as on the preceding time interval (t_N,t_{N+1}). In this case, it suffices to continue the approximate straight line $y=y_N(t,\lambda_N)$ to the intersection with the line $t_{N+2}=0$. The value of the function $y_{N+1}(t,\lambda_{N+1})$ at their intersection point will be one of the values of the prediction variable. In this case the controlling parameter equals zero, i.e., $\omega_m(t,\lambda_m)=0$. For any other t, taken in the interval $(t_{N+1}\le t\le t_{N+2})$, the appropriate value of the function $\omega_{N+1}(t,\lambda_{N+1})$ will differ from the case $\omega_m(t,\lambda_m)=0$. The maximal value of the unaccounted parameters influence function will correspond for $t=t_{m+1}$, i.e., $\omega_m(t_{m+1},\lambda_m)$. Choosing by desire the function $\omega_m(t,\lambda_m)$ arising in the preceding time interval (t_1,λ_N), and influencing by it beginning from the point t_{N+1} to the point t_{N+2}, the function $y_{N+1}(t,\lambda_{N+1})$ will be a numerical value of the expected (predictable) event on the further step t_{N+2}. Taking into account the fact that we can choose by desire the function $\omega_m(t,\lambda_m)$, then this function will represent the unaccounted parameters influence function and its appropriate function will be a controlling aim function.

It should be noted the fact that speaking about the functions $\omega_m(t,\lambda_m)$, we should understand their preliminarily calculated values at preceding stages. Therefore, in Eq. (27), we use the ready numerical value $\omega_m(t,\lambda_m)$ calculated by constructing the preceding piecewise-linear functions.

Thus, by means of action of unaccounted parameters influence function of the form $\omega_m(t,\lambda_m)$ or influence of their combinations from the end of the piecewise-linear function $[t_{N+1};y_N(t_{N+1},\lambda_N)]$ the piecewise-linear functions will emanate as fan a (Fig. 2). And the intersection of this series of functions and the straight line $t_{N+2}=0$ will give a series of values of the functions $y_{N+1,1}(t_{N+2,}\ l_{N+1})$, $y_{N+1,2}(t_{N+2,}\ l_{N+1})$, …, $y_{N+1,m}(t_{N+2,}\ l_{N+1})$.

And this series of values of the aim function will create domain of its change wherein there will be minimum and maximum of its value $[y_{N+1,1}(t_{N+2,}\ l_{N+1})]_{min}$, and $[y_{N+1,1}(t_{N+2,}\ l_{N+1}),]_{max}$. This range of the function $y_{N+1}(t,\lambda_m)$ will serve as a control domain of the process.

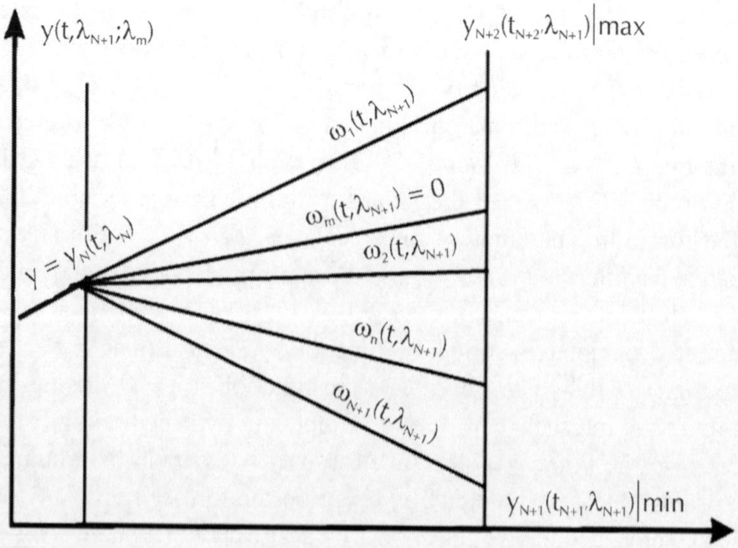

FIGURE 2 Determination of the range of predictable function.

4.4 METHOD OF NUMERICAL CALCULATION OF MODEL PLANE PROBLEMS WITH REGARD TO UNACCOUNTED PARAMETER INFLUENCE

Based around the developed model, we give a calculation method of concrete model problems of economics. For that we consider three types of model curves of the event which often met in the economic problems with downwards convexity, upwards convexity, and sinusoidal ones [1–4, 6].

4.4.1 *EXAMPLE 1*

Given a table of data $\{t_i, t_i, (t_i)\}$ describing some economic process (Table 1). By means of approximate methods, in particular by the method of least squares, represent this table in the form of five piecewise-linear functions (Fig. 3). It is seen from the picture that the approximate curve represents piecewise-linear curves with upwards convexity.

TABLE 1 Numerical construction of the 6-tier piecewise linear function with upwards convexity. Here $E_n = \frac{y_{n+1} - y_n}{t_{n+1} - t_n}$

n	t_n	$y_n(t_n)$	E_n	λ_n	$\omega_n(t,\lambda_n)$	$\omega_n(t_{n+1},t_n)$	$y_n(t,\lambda_n)$
1	1	2	4	0	0	0	$y_1(t)$ =4t–2
2	3	10	2	0.5	$0.5\,(1-\frac{3}{t})$	0.2857	$y_2(t,\lambda_2)$ =2t+4
3	7	18	1.2	0.58	$0.58\,(1-\frac{7}{t})$	0.2417	$y_3(t,\lambda_3)=$ =1.2t+9.6
4	12	24	0.5	0.7692	$0.7692\,(1-\frac{12}{t})$	0.2564	$y_4(t,\lambda_4)=$ =0.5t+17.9967
5	18	27	0.3333	0.7932	$0.7932\,(1-\frac{18}{t})$	0.1133	$y_5(t,\lambda_5)=$ =0.3332t+21
6	21	28					

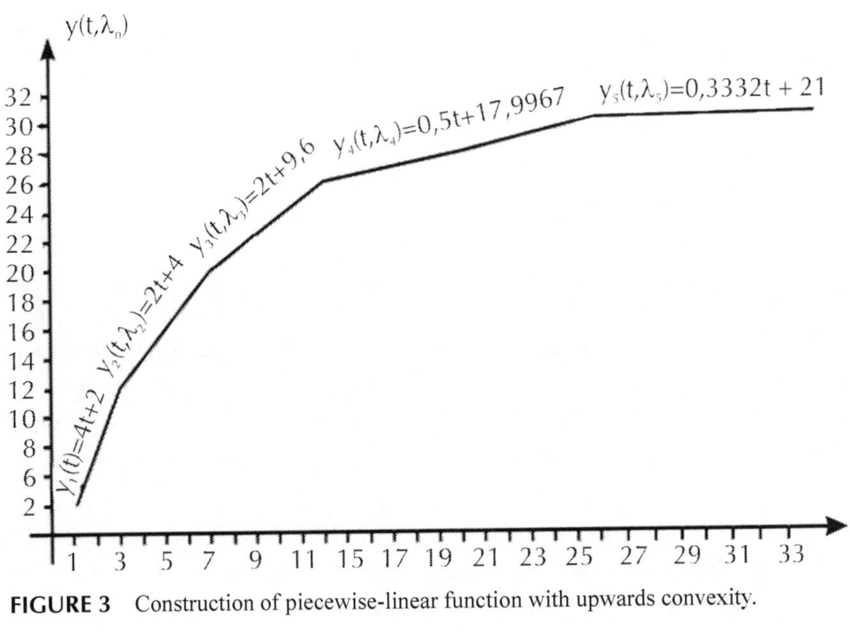

FIGURE 3 Construction of piecewise-linear function with upwards convexity.

The basic issue of the investigation is to construct the sixth piecewise-linear function on the subsequent time interval (t_1, t_6) knowing the characteristics of preceding five piecewise-linear functions given on the common time interval (t_6, t_7); secondly to give prediction recommendations of economic event and show a way for control of economic process by the characteristics of the six piecewise-linear function.

For that we apply the equation of the aim function Eqs. (27)–(29) to the problem under investigation. It will take the form:

$$y_6(t,\lambda_6,\lambda_m) = E_1 t[1-\lambda_m(1-\frac{t_6}{t})]\prod_{k=2}^{5}[1-\omega_k(t_{k+1},\lambda_k)]+ \\ + y_1(t_1) - E_1 t_1 \tag{30}$$

$$\lambda_m = \frac{\prod\limits_{\alpha=2}^{m-1\geq 2}[1-\omega_\alpha(t_{\alpha+1},\lambda_\alpha)] - \frac{E_m}{E_1}}{\prod\limits_{\alpha=2}^{m-1\geq 2}[1-\omega_\alpha(t_{\alpha+1},\lambda_\alpha)]} \tag{31}$$

$$\omega_m(t,\lambda_m) = \lambda_m(1-\frac{t_{N+1}}{t}) = \\ = \frac{\prod\limits_{\alpha=2}^{m-1\geq 2}[1-\omega_\alpha(t_{\alpha+1},\lambda_\alpha)] - \frac{E_m}{E_1}}{\prod\limits_{\alpha=2}^{m-1\geq 2}[1-\omega_\alpha(t_{\alpha+1},\lambda_\alpha)]}(1-\frac{t_{N+1}}{t}) \tag{32}$$

Thus, the function $y_6(t,\lambda_6;\lambda_m)$ depends on time t and arbitrary chosen parameter λ_m. And the parameter λ_m is the numerical value of the economic event obtained on the entire preceding time interval (t_1, t_6). To each numerical value of the parameter λ_m^* there will correspond its own aim function of the economic event $y_6(t,\lambda_6;\lambda_m^*)$. All these functions will emanate from one point $y_6(t_6;\lambda_6)$ in the form of a fan. For each value of λ_m^*, the intersection of the functions $y_6(t,\lambda_6;\lambda_m^*)$ and the straight line $t_7 = 0$ will determine both the range of the controlling aim function $y_6(t,\lambda_6;\lambda_m)$ and numerically predict the behavior of the process on the next time interval $t_7 \leq t \leq t_8$.

Give the values $E_1 = 4$, $t_1 = 1$, $y_1(t_1) = 2$, $t_6 = 21$ and making appropriate calculations (Table 1, Fig. 3) the equation of the sixth piecewise-linear straight line with upwards convexity curve will take the form:

$$y_6(t,\lambda_m) = 1{,}4286[1-\lambda_m(1-\frac{21}{t})]t-2 \tag{33}$$

By Eq. (33) we calculate for $t_7 = 24$ all the values of the unaccounted factors influence parameter $\lambda_m : \lambda_2 = 0{,}5;\ \ \lambda_3 = 0{,}58;\ \ \lambda_4 = 0{,}7682;\ \ \lambda_5 = 0{,}7932$ (of Tables 1 and 2; Figs. 3 and 4).

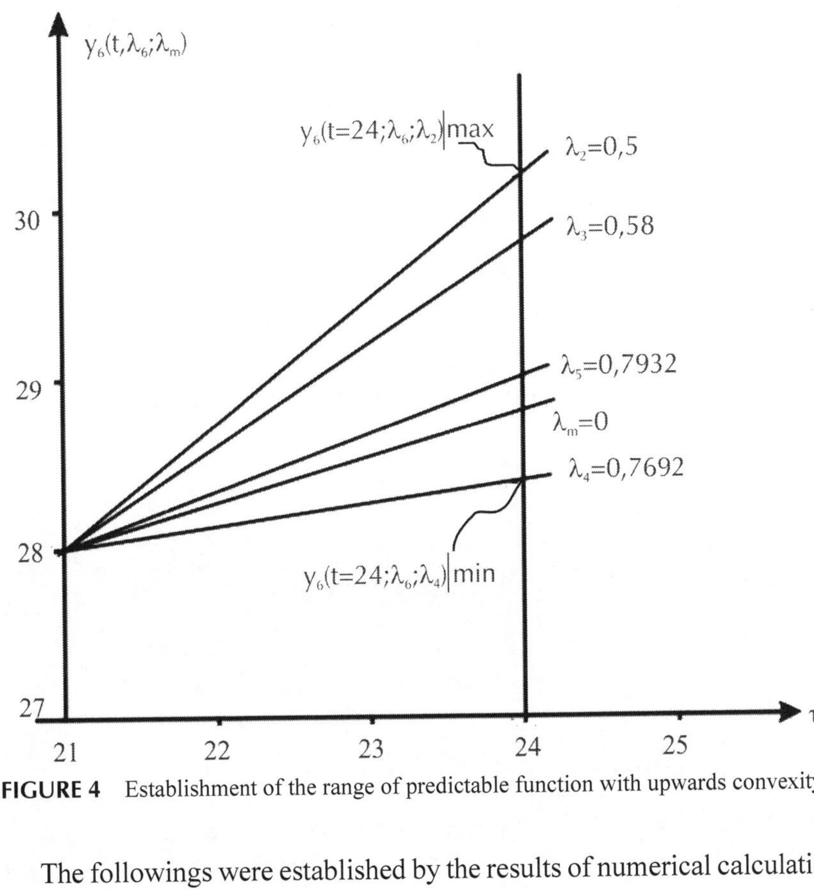

FIGURE 4 Establishment of the range of predictable function with upwards convexity.

The followings were established by the results of numerical calculation [1, 2, 4–6]:

1. According to the suggested model, the piecewise-linear continuous function $y = y_n(t, \lambda_n)$ with upwards convexity in the interval $1 \le t \le 21$ depending on the form of the first piecewise-linear function $y_1 = y_1(t)$ and all unaccounted parameters influence functions $\omega_n(t, \lambda_n)$ (Table 1, Fig. 3) was structured.

The quality property of the constructed aim function is in its chain character of relation between piecewise-linear functions by means of unaccounted factors. In practice, this means that the character of unaccounted factors of preceding stages of economic process will necessarily tell on quality value of the desired aim function on the subsequent stage.

2. The domain of possible change of the aim function of the event $y_6(t, \lambda_6; \lambda_m)$ on time interval $21 \le t \le 24$ depending on the control parameter λ_m (Table 2; Fig. 4) was numerically established.

TABLE 2 Numerical values of the range of change of the predictable function with upwards convexity.

$$y_6(t, \lambda_m) = 1{,}4286[1 - \lambda_m(1 - \frac{21}{t})]t - 2$$

$$y_6(t = 21; \lambda_6; \lambda_m = 0) = 28; \; y_5(t = 24; \lambda_5; \lambda_m = 0) = 29$$

m	λ_m	$y_6(t, \lambda_6; \lambda_m)$	$y_6(t_7 = 24, \lambda_6; \lambda_m)$
2	0.5	$y_6(t, \lambda_6; \lambda_2)$=0.7143t+13	30.1482
3	0.58	$y_6(t, \lambda_6; \lambda_3)$=0.6t+15.4	29.8.
4	0.7692	$y_6(t, \lambda_6; \lambda_4)$=0.3297$t$+21.076	28.62
5	0.7932	$y_6(t, \lambda_6; \lambda_5)$=0.2954$t$+21.7964	28.9888

$$28{,}2 \le y_6(t_7 = 24;, \lambda_6; \lambda_m) \le 30{,}1482 \qquad (34)$$

It was established that the expected prediction variables of the functions $y = y_6(t, \lambda_6; \lambda_m)$ on the range $21 \le t \le 24$ emanate from the point $t = 24$ as a fan and are determined by Eq. (34).

This means that the numerical value of the aim function expected on the subsequent stage of time of economic process, subject to unaccount of additional factors will equal 28. Under the unaccounted factor parameters it may be increased to 30.1482.

3. Domain of possible change of controls parameter λ_m in time range $1 \le t \le 24$:

$$0{,}5 \le \lambda_m \le 0{,}7932 \tag{35}$$

and domain of possible change of the control function $\omega_n(t_{n+1}, \lambda_n)$:

$$0{,}1133 \le \omega_n(t_{n+1}, \lambda_n) \le 0{,}2857 \tag{36}$$

were numerically established.

These ranges of the parameter and the function are the ranges of possible control of the process.

4.4.2 EXAMPLE 2

Table 3 describes some economic processes by means of the least squares method, which represents this table in the form of five piecewise-linear functions (Fig. 5).

TABLE 3 Numerical construction of the 6-tier piecewise linear function with downwards convexity.

Here $E_n = \dfrac{y_{n+1} - y_n}{t_{n+1} - t_n}$

n	t_n	$y_n(t_n)$	E_n	λ_n	$\omega_n(t, \lambda_n)$	$\omega_n(t_{n+1}, t_n)$	$y_n(t, \lambda_n)$
1	1	2	0,2				$y_1(t)$ =0.2t+ 1.8.
2	6	3	0,8.	–3	$-3^{(1-\frac{6}{t})}$	–1.3636	$y_2(t, \lambda_2) =$ =0.8t–1.8.

3	11	7	1.5555	−2.2905	$-2.2905\,(1-\frac{1}{t})$	−0.6649	$y_3(t,\lambda_3)=$ =1,5555t- −10.1099
4	15,5	14	3,2	−3.0659	$-3.0659\,(1-\frac{15,5}{t})$	−0.4259	$y_4(t,\lambda_4)=$ =3.2t- −35.5994
5	18	22	8	−6.1287	$-6.1287\,(1-\frac{18}{t})$	−0.3226	$y_5(t,\lambda_5)=$ =8t−122
6	19	30					

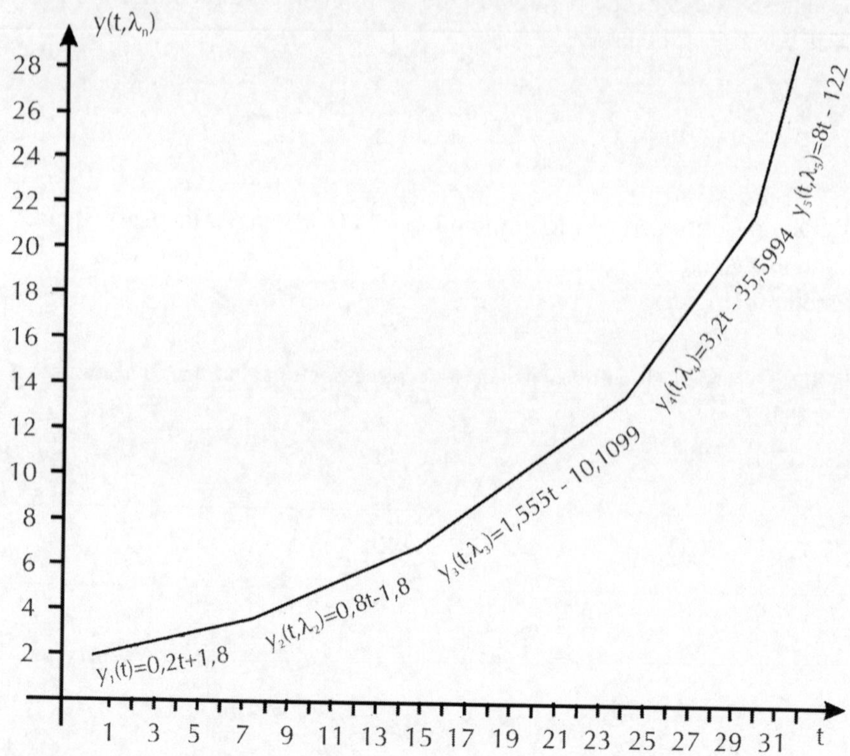

FIGURE 5 Construction of piecewise-linear function with downwards convexity.

It is seen from the Fig. 5 that the approximation curve is a piecewise-continuous function with downwards convexity.

The task of the investigation is to construct the sixth piecewise-linear function on the subsequent time interval (t_1, t_6), knowing the characteristics of preceding five piecewise-linear functions given on the common time interval (t_6, t_7), secondly, give prediction recommendations of economic process and show a way for economic process control by the characteristics of the sixth piecewise-linear function.

For solving the stated problem, apply the aim function Eqs. (27)–(29) to the problem under consideration.

It will take the form:

$$y_6(t, \lambda_6, \lambda_m) = E_1 t[1 - \lambda_m(1 - \frac{t_6}{t})]\prod_{k=2}^{5}[1 - \omega_k(t_{k+1}, \lambda_k)] + \\ + y_1(t_1) - E_1 t_1 \tag{37}$$

$$\lambda_m = \frac{\prod_{\alpha=2}^{m-1\geq 2}[1 - \omega_\alpha(t_{\alpha+1}, \lambda_\alpha)] - \frac{E_m}{E_1}}{\prod_{\alpha=2}^{m-1\geq 2}[1 - \omega_\alpha(t_{\alpha+1}, \lambda_\alpha)]} \tag{38}$$

$$\omega_m(t, \lambda_m) = \lambda_m(1 - \frac{t_{N+1}}{t}) =$$

$$= \frac{\prod_{\alpha=2}^{m-1\geq 2}[1 - \omega_\alpha(t_{\alpha+1}, \lambda_\alpha)] - \frac{E_m}{E_1}}{\prod_{\alpha=2}^{m-1\geq 2}[1 - \omega_\alpha(t_{\alpha+1}, \lambda_\alpha)]}(1 - \frac{t_{N+1}}{t}) \tag{39}$$

Thus, the aim function $y = y_6(t, \lambda_6; \lambda_m)$ determined on the subsequent interval depends on time t and arbitrarily chosen parameter λ_m. And the parameter λ_m are the numerical values of economic event, obtained on all the preceding time intervals. To each numerical value of the unaccounted factors influence parameter λ_m^* there will correspond its own economic process aim function $y = y_6(t, \lambda_6; \lambda_m^*)$.

All these functions will emanate from one point $y_6(t_6, \lambda_6)$ in the form of a fan. Intersection of the functions $y = y_6(t, \lambda_6; \lambda_m^*)$ and the straight line $t_7 = 0$ for each value of λ_m^* will determine the domain of controlling aim function $y = y_6(t, \lambda_6; \lambda_m)$ and allow to predict numerically the economic process behavior on the subsequent time interval $t_7 \le t \le t_8$.

Giving the values $E_1 = 0{,}2, t_1 = 1, y_1(t_1) = 2, t_6 = 19$ and making appropriate calculations (Table 3, Fig. 5), the equation for the sixth piecewise-linear function with downwards convexity will accept the form:

$$y_6(t, \lambda_m) = 1{,}4843[1 - \lambda_m(1 - \frac{19}{t})]t - 1{,}8 \tag{40}$$

By Eq. (40), the calculation was carried out at $t_7 = 24$ for all the values of the unaccounted factors influence parameter $\lambda_m : \lambda_2 = -3; \quad \lambda_3 = -2{,}2905;$ $\lambda_4 = -3{,}0659; \; \lambda_5 = -6{,}1287$ (Tables 3 and 4; Figs. 5 and 6).

TABLE 4 Numerical values of the range of change of the predictable function with downwards convexity.

$$y_6(t, \lambda_m) = 1{,}4843[1 - \lambda_m(1 - \frac{19}{t})]t - 1{,}8$$

$$y_6(t = 19; \lambda_6; \lambda_m = 0) = 30; \; y_5(t = 24; \lambda_5; \lambda_m = 0) = 70$$

m	λ_m	$y_6(t, \lambda_6; \lambda_m)$	$y_6(t = 24, \lambda_6; \lambda_m)$
2	–3	$y_6(t, \lambda_6; \lambda_2)$ =5.9372t–2.8051	59.6877
3	–2.2905	$y_6(t, \lambda_6; \lambda_3)$ =4.8891t–62.796	54.4227
4	–3.0659	$y_6(t, \lambda_6; \lambda_4)$ =6.035t–84.6636	60.1764
5	–6.1287	$y_6(t, \lambda_6; \lambda_5)$ =10.5811t–71.0398	82.9066

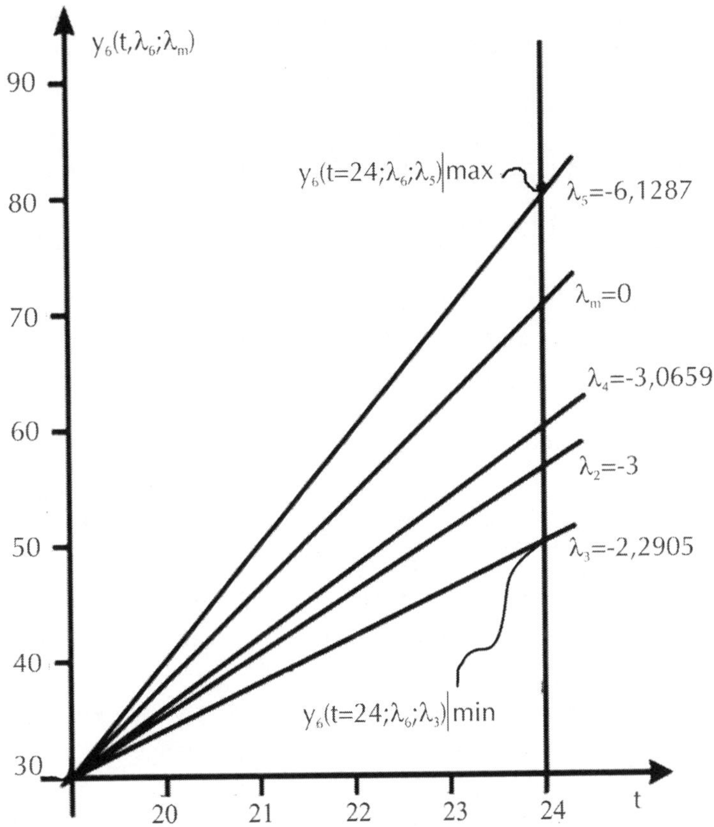

FIGURE 6 Establishment of the range of predictable function with downwards convexity.

The followings were established by the results of numerical calculation [1, 2, 4–6]:

1. According to the suggested model, a piecewise-linear continuous function $y = y_n(t, \lambda_n)$ with downwards convexity in time interval $1 \le t \le 19$ depending on the form of the first piecewise-linear function $y_1 = y_1(t)$ and all unaccounted parameters (factors) influence function $\omega_n(t, \lambda_n)$ (Table 3, Fig. 5) was constructed.
   The quality property of the constructed aim function is in its chain character of relation between piecewise-homogeneous linear functions by means of unaccounted factors. This means that unaccount-

ed factors of preceding stages will influence on quality value of the aim function on the subsequent stage.

2. The domain of possible change of the aim function of the event $y = y_6(t, \lambda_6; \lambda_n)$ on time interval $19 \le t \le 24$ depending on the control parameter λ_m (Table 4; Fig. 6) was numerically established.

$$59{,}6877 \le y_6(t_7 = 24, \lambda_6; \lambda_m) \le 82{,}9066 \tag{41}$$

It was established that for the case of a curve with downwards convexity, the expected predictable quantities of functions $y = y_6(t, \lambda_6; \lambda_m)$ on time interval $19 \le t \le 24$ emanate from the point $t = 19$in the form of a fan and are determined by Eq. (41).
This means the following. The numerical value of the aim function expected on the subsequent state will equal 30 subject to unaccount of additional factors. Under the influence of only unaccounted factor influence it may increase to 82.9066. This fact shows that at rational control of unaccounted factors parameters of the considered economic process, the described curve with downwards convexity may increase by 2,5 times.

3. The domain of change of the control parameter λ_m on time interval $1 \le t \le 24$:

$$-6{,}1287 \le \lambda_m \le -3 \tag{42}$$

and domain of change of the control function $\omega_n(t_{n+1}, t_n)$ of the event:

$$-1{,}3636 \le \omega_n(t_{n+1}, \lambda_n) \le -0{,}3226 \tag{43}$$

were numerically established

4.4.3 EXAMPLE 3

Consider the case when the curve is a piecewise-linear function of sinusoidal type (Table 5, Fig. 7).

TABLE 5 Numerical construction of sinusoidal type piecewise linear function.

n	t_n	$y_n(t_n)$	E_n	λ_n	$\omega_n(t,\lambda_n)$	$\omega_n(t_{n+1},t_n)$	$y_n(t,\lambda_n)$
1	1	2	4	0	0	0	$y_1(t) = 4t - 2$
2	2	6	1.3333	0.6667	$0.6667\ (1-\frac{2}{t})$	0,4	$y_2(t,\lambda_2) =$ =1.3332t +3.3336
3	5	10	−1	1.4167	$1.4167\ (1-\frac{5}{t})$	0.6296	$y_3(t,\lambda_3)$ =-t+15
4	9	6	0.6666	0.2502	$0.2502\ (1-\frac{9}{t})$	0.0625	$y_4(t,\lambda_4) =$ =0.6665t+0.0018
5	12	8.	3	−2.6006	$-2.6006\ (1-\frac{12}{t})$	−0.3715	$y_5(t,\lambda_5)$ =3t-28
66	14	14	7.5	−5.5628	$-5.5628\,(1-\frac{14}{t})$	−0.6953	$y_6(t,\lambda_6)$ =7,5t-91
7	16	29	−1	1.5163	$1.5163\ (1-\frac{16}{t})$	0.4615	$y_7(t,\lambda_7)$ =-t+45
8.	23	22	3	−1.8758	$-1.8758\ (1-\frac{23}{t})$	−0.1501	$y_8(t,\lambda_8)$ =3t-47
9	25	28	0				

Here $E_n = \dfrac{y_{n+1} - y_n}{t_{n+1} - t_n}$

According to the characteristics of preceding eight piecewise-linear functions given on the common time interval (t_1, t_9), construct the nineth piecewise-linear function determined on the subsequent time interval (t_9, t_{10}); by the characteristics of the constructed nineth piecewise-linear function give the prediction recommendations of economic event and show the way of economic process control.

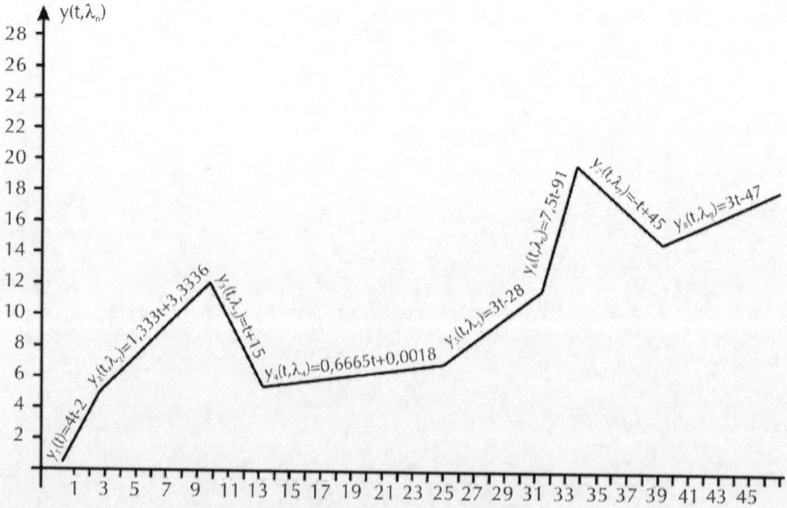

FIGURE 7 Construction of piecewise-linear function of sinusoidal type.

For solving the stated problem, apply the aim function Eq. (27)–(29) to the problem under investigation. It will take the form:

$$y_6(t,\lambda_8,\lambda_m) = E_1 t[1-\lambda_m(1-\frac{t_9}{t})]\prod_{k=2}^{8}[1-\omega_k(t_{k+1},\lambda_k)]+ \tag{44}$$

$$+y_1(t_1)-E_1t_1$$

$$\lambda_m = \frac{\prod\limits_{\alpha=2}^{m-1\geq 2}[1-\omega_\alpha(t_{\alpha+1},\lambda_\alpha)]-\dfrac{E_m}{E_1}}{\prod\limits_{\alpha=2}^{m-1\geq 2}[1-\omega_\alpha(t_{\alpha+1},\lambda_\alpha)]} \tag{45}$$

$$\omega_m(t,\lambda_m) = \lambda_m(1-\frac{t_{N+1}}{t}) =$$

$$= \frac{\prod\limits_{\alpha=2}^{m-1\geq 2}[1-\omega_\alpha(t_{\alpha+1},\lambda_\alpha)]-\dfrac{E_m}{E_1}}{\prod\limits_{\alpha=2}^{m-1\geq 2}[1-\omega_\alpha(t_{\alpha+1},\lambda_\alpha)]}(1-\frac{t_{N+1}}{t}) \tag{46}$$

Thus, the aim function $y = y_9(t, \lambda_8; \lambda_m)$ on the subsequent time interval is expressed depending on time t and arbitrarily chosen parameters λ_m. And the parameter λ_m are the numerical values of the economic process, obtained on all the preceding time interval (t_1, t_9). To each value of the unaccounted factors influence parameter λ_m^* there will correspond its own economic process aim function $y = y_9(t, \lambda_8; \lambda_m^*)$. All these functions will emanate from one point $y_9(t_9, \lambda_9)$ in the form of a fan. The intersection of the functions $y = y_9(t = 30, \lambda_9; \lambda_m^*)$ and the straight line $t_{10} = 30$ for each value of λ_m^* will determine the domain of change of the controlling aim function and allow to predict (prognose) numerically the behavior of the process on the next time interval (25 £ t £ 30).

Giving the values E_1 = four; $t_1 = 1$; $y_1(t_1) = 2$; $t_9 = 25$ and carrying out appropriate calculations (Table 5, Fig. 7) the equation for the nineth piecewise-linear function of sinusoidal type (Fig. 8) will take the form:

$$y_9(t, \lambda_m) = 1{,}2[1 - \lambda_m(1 - \frac{25}{t})]t - 2 \tag{47}$$

For the aim function $y = y(t, \lambda_9)$ of sinusoidal type, the numerical calculation was carried out by equation $t_{10} = 30$ for all the values of the control parameter λ_m (unaccounted, not planned factors influence parameter) (Table 6, Fig. 8): $\lambda_2 = 0{,}6667$; $\lambda_3 = 1{,}4167$; $\lambda_4 = 0{,}2502$, $\lambda_5 = 2{,}6006$; $\lambda_6 = 5{,}5628$; $\lambda_7 = 1{,}5163$; $\lambda_8 - 1{,}8758$.

TABLE 6 Numerical values of the range of change of the predictable function of the sinusoidal type.

$$y_9(t, \lambda_m) = 1{,}2[1 - \lambda_m(1 - \frac{25}{t})]t - 2$$

$$y_9(t = 25; \lambda_9; \lambda_m = 0) = 28; \; y_8(t = 30; \lambda_8; \lambda_m = 0) = 43$$

m	λ_m	$y_9(t, \lambda_9; \lambda_m)$	$y_9(t_{10} = 30, \lambda_9; \lambda_m)$
2	0.6667	$y_9(t, \lambda_9; \lambda_2)$=0.4t+18	30
3	1.4167	$y_9(t, \lambda_9; \lambda_3)$ =–0.5t+40.5	25.5

4	0.2502	$y_9(t,\lambda_9;\lambda_4)$=0.9t+five.5	34.07
5	–2.6006	$y_9(t,\lambda_9;\lambda_5)$=4.3207t–80	49.603
6	–5.5628	$y_9(t,\lambda_9;\lambda_6)$=7.8754t–168.884	67.378
7	1.5163	$y_9(t,\lambda_9;\lambda_7)$=–0.62t+43.489	24.889
8	–1.8758	$y_9(t,\lambda_9;\lambda_8)$=4.651t–88.274	51.256

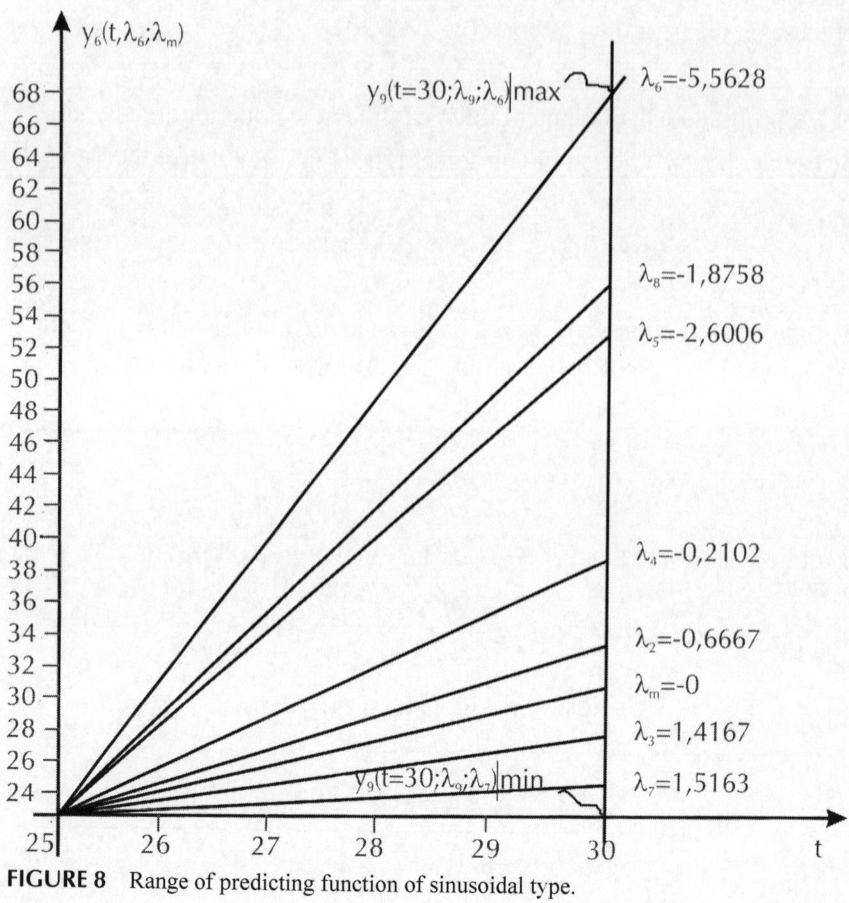

FIGURE 8 Range of predicting function of sinusoidal type.

The results of the numerical calculation established [1, 2, 4–6]:

(1) According to the suggested dynamic model, a piecewise-linear continuous function $y = y_n(t, \lambda_n)$ of sinusoidal type in time change interval $1 \leq t \leq 25$ depending on the form of the first piecewise-linear function $y = y_1(t)$ and all unaccounted factors influence function $\omega_n(t_{n+1}, \lambda_n)$ (Table 5, Fig. 7) was constructed.
Under sinusoidal character of the economic process, the peculiarity of the constructed aim function will be determined both by the character of its chain functional relation between piecewise-homogeneous linear functions and by the factor of its sign changeability.

(2) Domain of possible change of the aim function of the event's $y = y_9(t, \lambda_9; \lambda_m)$ on the time change interval 25 £ t £ 30 depending on the control parameters λ_m (Table 6; Fig. 8) was numerically determined

$$24{,}889 \leq y_9(t_{10} = 30, \lambda_9; \lambda_m) \leq 67{,}378 \tag{48}$$

This means that if we disregard the influence of additional factors, than the numerical value of the prediction quantity expected on the subsequent stage will equal 28. But by taking into account such factors, this number will increase by 2.3 times.

(3) It was established that the expected prediction values of the functions $y = y_9(t, \lambda_9; \lambda_m)$ on time change interval 25 £ t £ 30 will emanate as a fan from the point t = 25 and will be determined by Eq. (47). The domain of change of the control parameter λ_m in time change interval 25 £ t £ 30:

$$-5.5628 \leq \lambda_m \leq 1.5163 \tag{49}$$

and domain of change of the event control function $\omega_n(t_{n+1}, \lambda_n)$:

$$-0.6953 \leq \omega_n(t_{n+1}, \lambda_n) \leq 0.6296 \tag{50}$$

in time interval 25 £ t £ 30 were numerically established.

KEYWORDS

- **downwards convexity**
- **piecewise-homogeneous linear function**
- **piecewise-linear economic-mathematical models**
- **sinusoidal type**
- **upwards convexity**

REFERENCES

1. Aliyev Azad, G. *Development of dynamical model for economic process and its application in the field of industry of Azerbaijan.* Thesis for Phd, Baku, **2001,** 167.
2. Aliyev Azad, G. *Economic-mathematical methods and models with regard to incomplete information.* Ed. “Elm,” ISBN 5–8066–1487–5, Baku, **2002,** 288.
3. Aliyev Azad, G.; Ecer, F. *Tam olmayan bilqiler durumunda iktisadi matematik metodlar ve modeler,* Ed. NUI of Turkiye, ISBN 975-8062-1802, Nigde, 2004, 223.
4. Aliyev Azad, G. *Economic-mathematical methods and models in uncertainty conditions in finite-dimensional vector space.* Ed. NAS of Azerbaijan “Information technologies.” ISBN 978–9952-434-10-1, Baku, **2009,** 220.
5. Aliyev Azad, G. *Theoretical bases of economic-mathematical simulation at uncertainty conditions*, National Academy of Sciences of Azerbaijan, Baku, “Information Technologies,” ISBN 995221056-9, **2011,** 338.
6. Aliyev Azad, G. *Economical and mathematical models subject to complete information.* Ed. Lap Lambert Akademic Publishing. ISBN 978-3-659-30998-4, Berlin, **2013,** 316.
7. Aliyev Azad, G. Ekonomi Meselelerin Cozumune Iliskin Matematik Dinamik Bir Model. Dergi ‘Bilgi ve Toplum’. Turkiye, Istanbul, Milli Yayin, Nu 99-34-Y-0147, ISBN 975-478-120-5, Istanbul-**1999,** 83–106 p., (Turkish).
8. Aliyev Azad, G. Dynamical analysis method of production volume in industrial fields of Azerb. Republic. Dep. in AzNIINTI, *2627,* 44p, Baku-**1999,** Review col. of dep. papers, *1(16)*, 11 p. (Russian).
9. Aliyev Azad, G. Dynamical analysis method of production volume in oil and gas production and oil-chemical industries of Azerb. Rep., Dep. AzNIINTI, *2629,* 16p. Baku-**1999,** Review Col. of dep. Papers, *1(16),* 12 p. (Russian).
10. Aliyev Azad, G. Dynamical analysis method of wages fund in industrial fields of Azerbaijan. Dep. AzNIINTI, 2626, 89p. Baku-**1999,** Review Col. of dep. Papers, *1(16),* 11 p. (Russian).
11. Aliyev Azad, G. Dynamical analysis method of wages fund in oil and gas production and oil-chemical fields of industry of Azerbaijan. Dep. AzNIINTI, *2628,* 15p. Baku **1999,** Review col. of dep. papers, *1(16),* 11p. (Russian).

12. Aliyev Azad, G. On a dynamical model of investigation for economical problems. Trudy. Inst. Mat. i Mech. Acad. of Sci. of Azerbaijan. Baku, **1998,** *9,* 195–203, ISSN 0207–3188. (Russian).
13. Aliyev Azad, G. Features of construction of dynamic models in view of influence of the not discounted factors, "Strategic problems of economy of Azerbaijan", Theses of the reports of a republican scientific-practical conference, In the Azerbaijan State Oil Academy, Baku, **2002**, 81–84, 4 p. (Russian).
14. Aliyev Azad, G. Method of construction of economical-mathematical models, prediction and control of economical event when information is incomplete. Proc. of the II International scientific-practical conference of young Scientists, Kazakh National Technical University, 9965-585-50-4, Almaty, **2002,** 159–162, 4 p. (Russian).
15. Aliyev Azad, G. Application of mathematical methods in business activity. Methodical instructions. Publ. house of Azerbaijan Oil Academy. Baku, **2002,** 60 p. (Russian).
16. Aliyev Azad, G. Ismailova L.G., About one economic-mathematical model of management and forecasting of economic events in modern market conditions," Strategic problems of economy of Azerbaijan, "the Theses of the reports of a republican scientific – practical conference," In the Azerbaijan State Oil Academy, Baku, **2002,** 87–90, 4 p. (Russian).
17. Aliyev Azad, G. Method of construction of economical-mathematical models, prediction and control of economical event when information is incomplete, "Urgent problems of an economy of the transition period" (collection of The proceedings), Baku-2002, In the Azerbaijan State Oil Academy, 318–323, 6 p. (Russian).
18. Aliyev Azad, G., Model of influence of the external factors on course of economic process. Materials of 4-th international conference Economy and safety of ability to Live", Azerbaijan Sumgayit State University, Sumgayit, **2002,** 172–173, 2 p. (Russian).
19. Aliyev Azad, G. Technique of construction of economic-mathematical models, forecasting and management of economic event in conditions of the incomplete information, "Urgent problems of an economy of the transition period" (collection of the proceedings), In the Azerbaijan State Oil Academy, Baku-**2002,** 318–323, 6 p. (Russian).
20. Aliyev Azad, G. Fundamental scheme of construction of the economical-mathematical models for deciding of problem of the determination of affectivity of the production of electrical energy from different types of fuel resources, The problems of using of economical potential by enterprises in transitional period. Baku, **2003,** The Azerbaijan State Oil Academy, UOTJ 334.012.23, 143–148, 6 p. (Russian).
21. Aliyev Azad, G. Some aspects of regression analyze and forecasting of electric energy prime cost on employing of different fuel types is proposed in the work, News of Azerbaijan High Technical Educational Institutions, Baku, **2004,** The Azerbaijan State Oil Academy, *5(27),* ISSN 1609–1620, 71–77, 7 p, (Russian).
22. Aliyev Azad, G., The application of mathematical methods in organization of the external economical activity of the enterprise, Methodical instructions, Publ. The Azerbaijan State Oil Academy, Baku, **2004**. (Russian).
23. Aliyev Azad, G. The model of the definition of required profit and price for accumulation of the planning volume of private financial resources in careless conditions, Journal "Knowledge" "Education" Society of The Azerbaijan Republic, Baku, **2005,** Social Sciences 3–4, ISBN 1683–7649, 27–31, 5 p. (Russian).

24. Aliyev Azad, G., Some aspects of regressive analyze and prognosis of efficient of the labor in AOGE, News of Azerbaijan High Technical Educational Institutions, The Azerbaijan State Oil Academy, Baku, **2006,** *1(41),* ISSN 1609–1620, 64–69, 6 p. (Russian).
25. Aliyev Azad, G. The features of using integrated operational schedule in realization repair works. The proceedings of Azerbaijan scientific and research Institute of Economics and Organization of Agricultural farming, Baku, **2006,** *1,* 34–46, 13 p. (Russian).

CHAPTER 5

BASES OF SOFTWARE FOR COMPUTER SIMULATION AND MULTIVARIANT PREDICTION OF ECONOMIC EVEN AT UNCERTAINTY CONDITIONS ON THE BASE OF N-COMPONENT PIECEWISE-LINEAR ECONOMIC-MATHEMATICAL MODELS IN M-DIMENSIONAL VECTOR SPACE

CONTENTS

5.1 ON URGENCY OF DEVELOPMENT OF SOFTWARE FOR COMPUTER SIMULATION AND MULTIVARIANT PREDICTION OF ECONOMIC EVENT AT UNCERTAINTY CONDITION ON THE BASE OF *N*-COMPONENT PIECEWISE-LINEAR ECONOMIC-MATHEMATICAL MODELS IN *M*-DIMENSIONAL VECTOR SPACE

Development of modern society is characterized by the increase of technical level, complication of organizational structure of production, intensification of social division of labor, making high demands on planning and economic management methods-different optimization models and optimization methods based on the use of mathematical simulation find effective application by solving practical operational problems.

Today, newest achievements of mathematics and up-to-date calculating engineering find wider application in economic investigations and planning. According to the basic conditions, the simulation process stages acquire their specific character.

5.1.1 STATEMENT OF ECONOMIC PROBLEM AND ITS QUALITY ANALYSIS

The given stage means explicit formulization of the problem's essence, accepted assumptions. This stage includes distinction of the most important features and properties of the modeled object and its abstraction from the secondary ones; study of the structure and basic dependences connecting its elements; formulation of conjectures (even if preliminary ones), explaining the behavior and development of the object.

5.1.2 CONSTRUCTION OF MATHEMATICAL MODEL

This stage is the stage of formalization of an economic problem, its expressions in the form of concrete mathematical dependences and relations (functions, equations, inequalities and etc.). Usually, at first the basic construction (type) of a mathematical model is determined, then the details of this construction (concrete list of variables and parameters, connection forms) are specified. It is incorrect to assume that the more facts takes into account a model, the best it "works" and gives best results. We can say the same on such characteristics of complexity of the model as the used forms of mathematical dependences (linear and nonlinear), accounting of accidental nature and uncertainty factors and so on. Superfluous complexity and awkwardness of the model makes difficult the investigation process. It is necessary to take into account not only real possibilities of information and software but also to compare simulation expenditure with the obtained efficiency (by increasing the complexity of the model, increase of expenditures may exceed the efficiency increase).

Intercomparison of two systems of scientific knowledge's i.e., economic and mathematical ones are realized in the process of construction of the model. It is natural to try to get a model belonging to the well-studied class of mathematical problems. Often it is succeeded to do it by simplifying initial premises of the model that don't distort the essential features of the modeled object. However, the situation when the formalization of the economic problem reduces to the mathematical structure unknown earlier is also possible.

5.1.3 MATHEMATICAL ANALYSIS OF THE MODEL

The goal of this stage is elucidation of general properties of the model. Here truly mathematical investigation methods are used. The most important moment is to prove the existence of solutions in the formulated model (existence theorem). If it turns out well to prove that a mathematical problem has no solution, then the necessity in the subsequent work on the initial variant of the model falls away then either the statement of the economic problem or the ways of its mathematical formalization should be corrected. By analytic investigation of the model such questions as for example, if the solution is unique, which variables (unknown ones) may appear in the solution, what relations will be between them, in what limits and under which initial conditions they change, what tendencies of their change and etc. Analytic investigation of the model compared with empiric (numerical) one has the advantage that the obtained conclusions remain valid for different concrete values of external and internal parameters of the model.

As the economic-mathematical simulation develops and gets complicated, its separate stages are isolated into specialized investigation fields, the difference between theoretical-analytical and applied models increases, differentiation of models by the levels of abstraction and idealization happens.

Theory of mathematical analysis of economics models has been developed into a special branch of contemporary mathematics to mathematical economics. The models studied within mathematical economics loose direct connection with economic reality; they exceptionally deal with idealized economic objects and situations. By constructing such models, the chief principle is not so much approximation to reality as to obtain a possible great number of analytic results by means of mathematical proofs. The value of these models for economic theory and practice is that they serve as a theoretical basis for applied type models.

Preparation and processing of economic information and development of software of economic problems (creation of data base and information banks, program of computer-aided construction of models and program service for economists-users) become independent fields of investigations.

5.1.4. PREPARATION OF INPUT INFORMATION

Simulation presents rigid requirements to the information system. At the same time, real possibilities of information receipt restricts the choice of models intended for practical use. Not only principal possibility of information preparation (for certain periods) but also expenditures for preparation of appropriate information areas is taken into account. These expenditures should not exceed the efficiency from the use of additional information.

The methods of probability theory, theoretical and mathematical statistics are widely used in the course of preparation of information. Under system economic-mathematical simulation the initial information used in one model is the result of functioning of other models.

5.1.5 NUMERICAL CALCULATION

In this stage, the algorithms for numerical solution of the problem are worked out, the programs in economic-mathematical models are composed and calculations are conducted. Difficulties of this stage are stipulated first of all by the great size of economical problems, by necessity of processing of considerable information areas. Usually, calculations on economic-mathematical model are of multivariant character. Owing to high speed of contemporary ECM we can conduct numerous "model" experiments studying "behavior" of the model under different changes of some conditions. The investigations conducted by numerical methods may essentially complement the results of analytic investigation, and for a lot of models it is a uniquely realizable one.

5.1.6 ANALYSIS OF NUMERICAL RESULTS AND THEIR APPLICATION

On the final stage their arises a question on correctness and completeness of simulation, on degree of practical applicability of the latters. Mathematical verification methods may elucidate incorrect constructions of the model and by the same token contract the class of potentially tame models. Informal analysis of theoretical conclusion and numerical results obtained by means of the model, their comparison by the available knowledge's and facts of reality also allow to reveal the short-comings of the economic problem statement, constructed mathematical model and its software.

Introduction of computer-aided systems of economic information processing allows to lower essentially the expenditures connected with data processing, to increase labor productivity of the labor of the workers in the field of economics, improve relations between different subdivisions of enterprises.

At present, there is a great mass of software intended for application in the field of economics, but regretfully, often it is necessary to "adjust" the ready made software under individual features of the enterprise even if these programs stood the test by time.

In Chapters 2–4 of this monograph, a general theory of construction of n-component piecewise-linear economic-mathematical models at uncertainty conditions is developed, a new method of multi variant prediction of economic event in m-dimensional vector space is suggested.

However, the arising difficulties of calculating character require the creation of special software for computer programming and creation of an action algorithm for economic processes at uncertainty conditions in finite-dimensional vector space.

In this connection, in Refs. [1–4], by means of 2-component piecewise linear economic-mathematical models with regard to unaccounted factors a special program is developed for computer modeling for numerical construction and definition of multivariant prediction quantities of economic event in many-dimensional vector space, in particular, in 2-, 3- and 4-dimensional vector spaces. The scientific results obtained in these works compose necessary theoretical and calculation instrument for creating a principally new, perspective software for computer modeling by constructing and multivariant prediction of economic state by means of piecewise linear economic-mathematical models with regard to unaccounted factors influence in m-dimensional vector space.

In this chapter, the developed software algorithms for constructing 2-component piecewise-linear model and for multivariant prediction of economic event at uncertainty conditions in m-dimensional vector space will be stated on the base of the *Matlab* program, and a number of numerical examples will be given.

A packet of programs will be suggested, a numerical analysis of multivariant prediction of economic event at uncertainty conditions will be suggested.

5.2 DEVELOPMENT OF SOFTWARE FOR COMPUTER MODELING AND MULTIVARIANT PREDICTION OF ECONOMIC EVENT AT UNCERTAINTY CONDITIONS ON THE BASE OF 2-COMPONENT PIECEWISE-LINEAR ECONOMIC-MATHEMATICAL MODELS IN *M*-DIMENSIONAL VECTOR SPACE

5.2.1 ACTIONS ALGORITHM FOR COMPUTER MODELING BY CONSTRUCTING 2-COMPONENT PIECEWISE-LINEAR ECONOMIC-MATHEMATICAL MODELS

In this section, on the basis of the *Matlab* program We suggest an algorithm and numerical calculation method for numerical construction of

2-component piecewise-linear economic-mathematical models in *m*-dimensional vector space.

It should be noted that the *Matlab* program has its restrictive properties that compels us to introduce some additional denotation and adhere to certain proper sequence in calculation operations.

According to the suggested theory [5, 6] for the case of 2-component piecewise-linear vector function in *m*-dimensional vector space we write the min equations and mathematical expressions that are subjected to numerical programming.

Let in *m*-dimensional vector space R_m a statistical table describing some economic process in the form of points (vectors) set $\{\vec{a}_n\}$ be given. Let these points be represented in the form of adjacent seven-component piecewise linear vector equation of the form [4, 7, 8, 9]:

$$\vec{z}_1 = \vec{a}_1 + \mu_1(\vec{a}_2 - \vec{a}_1) \tag{1}$$

$$\vec{z}_2 = \vec{z}_1^{k_1} + \mu_2(\vec{a}_3 - \vec{z}_1^{k_1}) \tag{2}$$

where $\vec{z}_1 = \vec{z}_1(z_{11}, z_{12}, z_{13}, \ldots z_{1m})$ and $\vec{z}_2 = \vec{z}_2(z_{21}, z_{22}, z_{23}, \ldots z_{2m})$ are the equations of the first and second piecewise-linear straight lines in m-dimensional vector space. The vectors $\vec{a}_1(a_{11}, a_{12}, a_{13}, \ldots, a_{1m})$, $\vec{a}_2 = \vec{a}_2(a_{21}, a_{22}, a_{23}, \ldots, a_{2m})$ and $\vec{a}_3 = \vec{a}_3(a_{31}, a_{32}, a_{33}, \ldots, a_{3m})$ are the given points (vectors) in *m*-dimensional space, of the form:

$$\vec{a}_1 = a_{11}\vec{i}_1 + a_{12}\vec{i}_2 + a_{13}\vec{i}_3 + \ldots + a_{1m}\vec{i}_m,$$

$$\vec{a}_2 = a_{21}\vec{i}_1 + a_{22}\vec{i}_2 + a_{23}\vec{i}_3 + \ldots a_{2m}\vec{i}_m \tag{3}$$

$$\vec{a}_3 = a_{31}\vec{i}_1 + a_{32}\vec{i}_2 + a_{33}\vec{i}_3 + \ldots + a_{3m}\vec{i}_m$$

Here $\mu_1^{k_1} = \mu_1^*$ and $\mu_2^{k_2} = \mu_2^*$ are arbitrary parameters, $\vec{z}_1^{k_1}$ is the intersection point of the straight lines $\vec{z}_1$ and $\vec{z}_2$.

The goal of the investigation is the following. Giving the approximate points $\vec{a}_1$, $\vec{a}_2$, $\vec{a}_3$ and also the values of the parameters $\mu_1^{k_1} = \mu_1^{*}$ and $\mu_2^{k_2} = \mu_2^{*}$ to develop a computer calculation algorithm of the following equations and mathematical expressions in *m*–dimensional vector space:

$$\vec{z}_1^{k_1} = \vec{a}_1 + \mu_1^{k_1}(\vec{a}_2 - \vec{a}_1)$$

$$\mu_1^{k_2} = \mu_1^{k_1} + \mu_2^{k_2}\frac{(\vec{a}_3 - z_1^{k_1})^2}{(\vec{a}_3 - z_1^{k_1})(\vec{a}_2 - \vec{a}_1)}$$

$$\vec{z}_1^{k_2} = \vec{a}_1 + \mu_1^{k_2}(\vec{a}_2 - \vec{a}_1)$$

$$\vec{z}_2^{k_2} = \vec{z}_1^{k_1} + (\mu_1^{k_2} - \mu_1^{k_1})\frac{(\vec{a}_3 - \vec{z}_1^{k_1})(\vec{a}_3 - a_1)}{(\vec{a}_3 - \vec{z}_1^{k_1})^2}(\vec{a}_3 - \vec{z}_1^{k_1})$$

$$cos\alpha_{1,2} = \frac{(\vec{z}_1^{k_2} - \vec{z}_1^{k_1})(\vec{z}_1^{k_2} - \vec{z}_1^{k_1})}{\left|\vec{z}_1^{k_2} - \vec{z}_1^{k_1}\right|\left|\vec{z}_1^{k_2} - \vec{z}_1^{k_1}\right|}$$

$$A = (\mu_1^{k_1} - \mu_1^{k_2})\frac{\left|\vec{a}_2 - \vec{a}_1\right|\left|\vec{z}_1^{k_1} - \vec{a}_1\right|}{\vec{z}_1^{k_2}(\vec{z}_1^{k_1} - \vec{a}_1)}$$

$$\lambda_2 = \frac{\mu_2^{k_2}}{\mu_1^{k_1} - \mu_2^{k_2}} \cdot \frac{\left|\vec{z}_1^{k_2} - \vec{z}_1^{k_1}\right|\left|\vec{a}_3 - \vec{z}_1^{k_1}\right|}{\vec{z}_1^{k_2}(\vec{z}_1^{k_2} - \vec{z}_1^{k_1})} \cdot \frac{\vec{z}_1^{k_2}(\vec{z}_1^{k_1} - \vec{a}_1)}{\left|\vec{a}_2 - \vec{a}_1\right|\left|\vec{z}_1^{k_1} - \vec{a}_1\right|}$$

$$\omega_2(\lambda_2, \alpha_{1,2}) = \lambda_2 cos\alpha_{1,2}$$

$$\vec{z}_2 = \vec{z}_1^{k_2}\{1 + A[1 + \omega_2(\lambda_2, \alpha_{1,2})]\}$$

$$\vec{z}_1 = \vec{a}_1 + \mu_1(\vec{a}_2 - \vec{a}_1)$$

$$A(\mu_1) = (\mu_1^{k_1} - \mu_1)\frac{|\vec{a}_2 - \vec{a}_1||\vec{z}_1^{k_1} - \vec{a}_1|}{\vec{z}_1(\vec{z}_1^{k_1} - \vec{a}_1)}$$

$$A(\mu_1) = (\mu_1^{k_1} - \mu_1)\frac{|\vec{a}_2 - \vec{a}_1||\vec{z}_1^{k_1} - \vec{a}_1|}{\vec{z}_1(\vec{z}_1^{k_1} - \vec{a}_1)}$$

$$\lambda_2(\mu_1) = \frac{\mu_2}{\mu_1^{k_1} - \mu_1} \cdot \frac{|\vec{z}_1 - \vec{z}_1^{k_1}||\vec{a}_3 - \vec{z}_1^{k_1}|}{\vec{z}_1(\vec{z}_1 - \vec{z}_1^{k_1})} \cdot \frac{\vec{z}_1(\vec{z}_1^{k_1} - \vec{a}_1)}{|\vec{a}_2 - \vec{a}_1||\vec{z}_1^{k_1} - \vec{a}_1|}$$

$$\omega_2(\mu_1) = \lambda_2(\mu_1)Cos\alpha_{1,2}$$

$$\vec{z}_2(\mu_1) = \vec{z}_1\{1 + A(\mu_1)[1 + \omega_2(\mu_1)]\} \quad (4)$$

Introduce the following denotation:

$$\begin{aligned}
&\vec{a}_1 \to a1;\ \vec{a}_2 \to a2;\ \vec{a}_3 \to a3;\ \mu_1 \to m1;\ \mu_1^{k_1} \to m1k1;\\
&\mu_1^{k_2} \to m1k2,\ \vec{z}_1^{k_1} \to z1k1,\ \vec{z}_1^{k_2} \to z1k2;\quad \vec{z}_2^{k_2} \to z2k2;\\
&\vec{z}_1 \to z1;\ \mu_2 \to m2;\ \mu_2^{k_2} \to m2k2,\ A(\mu_1) \to Am1,\\
&\lambda_2 \to La2,\ \lambda_2(\mu_1) \to La2m1;\\
&\omega_2(\lambda_2, \alpha_{12}) = \omega_2(\mu_1) \to w2m1,\ \vec{z}_2 \to z2,\ \vec{z}_2(\mu_1) \to z2m1.
\end{aligned} \quad (5)$$

Using the introduced denotation Eq. (5), for the system Eq. (4) compose a program for numerical construction of 2-component piecewise-linear economic-mathematical models with regard to unaccounted factors influence in *m*-dimensional vector space in the *Matlab* program in the following form [4, 7–9]:

A_1=[a_{11} a_{12} a_{13} …a_1 *m*]
a_2=[a_{21} a_{22} a_{23} …a_2 *m*]

$a_3=[a_{31}\ a_{32}\ a_{33}\ \ldots a_3\ m]$

$m_1k_1=(m_1)*$

$m_2k_2=(m_2)*$

for $m_1=J_1: J_2:J_3$

$z_1k_1=a_1+m_1k_1*(a_2-a_1);$

$m_1k_2=m_1k_1+m_2k_2*((a_3-z_1k_1)*(a_3-z_1k_1)')/((a_3-z_1k_1)*(a_2-a_1)');$

$z_1k_2=a_1+m_1k_2*(a_2-a_1);$

$z_2k_2=z_1k_1+(m_1k_2-m_1k_1)*((a_3-z_1k_1)*(a_2-a_1)')/((a_3-z_1k_1)*(a_3-z_1k_1)')*(a_3-z_1k_1);$

$cosa_{12}=((z_1k_2-z_1k_1)*(z_2k_2-z_1k_1)')/(sqrt((z_1k_2-z_1k_1)*(z_1k_2-z_1k_1)')*sqrt((z_2k_2-z_1k_1)*(z_2k_2-z_1k_1)'))$

$A=(m_1k_1-m_1k_2)*(sqrt((a_2-a_1)*(a_2-a_1)')*sqrt((z_1k_1-a_1)*(z_1k_1-a_1)'))/(z_1k_2*(z_1k_1-a_1)');$

$p_1=m_2k_2/(m_1k_1-m_1k_2);$

$p_2=(sqrt((z_1k_2-z_1k_1)*(z_1k_2-z_1k_1)')*sqrt((a_3-z_1k_1)*(a_3-z_1k_1)'))/(z_1k_2*(z_1k_2-z_1k_1)');$

$p_3=(z_1k_2*(z_1k_1-a_1)')/(sqrt((a_2-a_1)*(a_2-a_1)')*sqrt((z_1k_1-a_1)*(z_1k_1-a_1)'));$

$La_2=p_1*p_2*p_3;$

$w_2=La_2*cosa_{12};$

$z_2=z_1k_2*(1+A*(1+w_2));$

$z_1=a_1+m_1*(a_2-a_1)$

$m_2=(m_1-m_1k_1)*(((a_3-z_1k_1)*(a_2-a_1)')/((a_3-z_1k_1)*(a_3-z_1k_1)'))$

Am_1=(m_1k_1–m_1)*(sqrt((a_2–a_1)*(a_2–a_1)'))*sqrt((z_1k–a_1)*(z_1k_1–a_1)'))/(z_1*(z_1k_1–a_1)')

p_1 m_1=m_2/(m_1k_1–m_1);

p_2 m_1=(sqrt((z_1–z_1k_1)*(z_1–z_1k_1)')*sqrt((a_3–z_1k_1)*(a_3–z_1k_1)'))/(z_1*(z_1–z_1k_1)');

p_3 m_1=(z_1*(z_1k_1–a_1)')/(sqrt((a_2–a_1)*($a_{2_}$1)')*sqrt((z_1k_1–a_1)*(z_1k_1–a_1)'));

La_2 m_1=p_1 m_1*p_2 m_1*p_3 m_1;

w_2 m_1=La_2 m_1*$cosa_{12}$

z_2 m_1=z_1*(1+Am_1*(1+w_2 m_1)) end (6)

5.2.2 ALGORITHM OF MULTIVARIANT COMPUTER MODELING OF PREDICTION VARIABLES OF ECONOMIC EVENT ON THE BASE OF 2-COMPONENT PIECEWISE-LINEAR ECONOMIC-MATHEMATICAL MODELS

In this section, we suggest a software algorithm for multivariant prediction of economic event at uncertainty conditions on the base of 2-component piecewise-linear economic-mathematical model in *m*-dimensional vector space [4, 9–11].

For the case of 2-component piecewise-linear vector function at uncertainty conditions in *m*-dimensional space on the base of the *Matlab* program we represent an algorithm and numerical program for multivariant prediction of economic event.

According to the theory [4, 9–11] for the case of 2-component piecewise-linear vector-function at uncertainty conditions in *m*-dimensional vector space we have the following equations and relations for multivariant prediction of the economic event:

$$\vec{z}_1^{k_1} = \vec{a}_1 + \mu_1^{k_1}(\vec{a}_2 - \vec{a}_1)$$

$$\mu_1^{k_2} = \mu_1^{k_1} + \mu_2^{k_2}\frac{(\vec{a}_3 - \vec{z}_1^{k_1})^2}{(\vec{a}_3 - \vec{z}_1^{k_1})(\vec{a}_2 - \vec{a}_1)}$$

$$\vec{z}_1^{k_2} = \vec{a}_1 + \mu_1^{k_2}(\vec{a}_2 - \vec{a}_1)$$

$$\vec{z}_2^{k_2} = \vec{z}_1^{k_1} + (\mu_1^{k_2} - \mu_1^{k_1})\frac{(\vec{a}_3 - \vec{z}_1^{k_1})(\vec{a}_2 - \vec{a}_1)}{(\vec{a}_3 - \vec{z}_1^{k_1})^2}(\vec{a}_3 - \vec{z}_1^{k_1})$$

$$cos\alpha_{1,2} = \frac{(\vec{z}_1^{k_2} - \vec{z}_1^{k_1})(\vec{z}_2^{k_2} - \vec{z}_1^{k_1})}{\left|\vec{z}_1^{k_2} - \vec{z}_1^{k_1}\right| \cdot \left|\vec{z}_2^{k_2} - \vec{z}_1^{k_1}\right|}$$

$$A(\mu_1^{k_2}) = A = (\mu_1^{k_1} - \mu_1^{k_2})\frac{\left|\vec{a}_2 - \vec{a}_1\right|\left|\vec{z}_1^{k_1} - \vec{a}_1\right|}{\vec{z}_1^{k_2}(\vec{z}_1^{k_1} - \vec{a}_1)}$$

$$\lambda_2(\mu_1^{k_2}) = \lambda_2^{k_2} = \frac{\mu_2^{k_2}}{\mu_1^{k_1} - \mu_1^{k_2}} \cdot \frac{\left|\vec{z}_1^{k_2} - \vec{z}_1^{k_1}\right|\left|\vec{a}_3 - \vec{z}_1^{k_1}\right|}{\vec{z}_1^{k_2}(\vec{z}_1^{k_2} - \vec{z}_1^{k_1})} \cdot \frac{\vec{z}_1^{k_2}(\vec{z}_1^{k_1} - \vec{a}_1)}{\left|\vec{a}_2 - \vec{a}_1\right|\left|\vec{z}_1^{k_1} - \vec{a}_1\right|}$$

$$\omega_2(\lambda_2^{k_2}, \alpha_{1,2}) = \lambda_2^{k_2} cos\alpha_{1,2}$$

$$\vec{z}_2(\mu_1^{k_2}) = \vec{z}_2 = \vec{z}_1^{k_2}\{1 + A\,[1 + \omega_2(\lambda_2^{k_2}, \alpha_{1,2})]\}$$

$$\vec{z}_1(\mu_1) = \vec{z}_1 = \vec{a}_1 + \mu_1(\vec{a}_2 - \vec{a}_1)$$

$$\mu_2 = (\mu_1 - \mu_1^{k_1}) \cdot \frac{(\vec{a}_3 - \vec{z}_1^{k_1})(\vec{a}_2 - \vec{a}_1)}{(\vec{a}_3 - \vec{z}_1^{k_1})^2}$$

$$A(\mu_1) = (\mu_1^{k_1} - \mu_1)\frac{\left|\vec{a}_2 - \vec{a}_1\right|\left|\vec{z}_1^{k_1} - \vec{a}_1\right|}{\vec{z}_1(\vec{z}_1^{k_1} - \vec{a}_1)}$$

$$\lambda_2(\mu_1) = \frac{\mu_2}{\mu_1^{k_1} - \mu_1} \cdot \frac{\left|\vec{z}_1 - \vec{z}_1^{k_1}\right|\left|\vec{a}_3 - \vec{z}_1^{k_1}\right|}{\vec{z}_1(\vec{z}_1 - \vec{z}_1^{k_1})} \cdot \frac{\vec{z}_1(\vec{z}_1^{k_1} - \vec{a}_1)}{\left|\vec{a}_2 - \vec{a}_1\right|\left|\vec{z}_1^{k_1} - \vec{a}_1\right|}$$

$$\omega_2(\lambda_2(\mu_1),\alpha_{12}) = \omega_2(\mu_1) = \lambda_2(\mu_1)cos\alpha_{1,2}$$

$$\vec{a}_4 = \sum_{j=1}^{m} a_{4j}\vec{i}_j$$

$$a_{42}(1) = (\vec{a}_4)_2 = z_{22}^{k_2} + \frac{a_{22} - a_{12}}{a_{21} - a_{11}}(a_{41}(1) - z_{21}^{k_2})$$

$$a_{43}(1) = (\vec{a}_4)_3 = z_{23}^{k_2} + \frac{a_{23} - a_{13}}{a_{21} - a_{11}}(a_{41}(1) - z_{21}^{k_2})$$

$$a_{44}(1) = (\vec{a}_4)_4 = z_{24}^{k_2} + \frac{a_{24} - a_{14}}{a_{21} - a_{11}}(a_{41}(1) - z_{21}^{k_2})$$

$$a_{4m}(1) = (\vec{a}_4)_m = z_{2m}^{k_2} + \frac{a_{2m} - a_{1m}}{a_{21} - a_{11}}(a_{41}(1) - z_{21}^{k_2})$$

$$q_1 = \frac{(\vec{a}_2 - \vec{a}_1)(\vec{a}_3 - \vec{z}_1^{k_1})}{(\vec{a}_3 - \vec{z}_1^{k_1})^2}$$

$$q_2 = \frac{(\vec{a}_3 - \vec{z}_1^{k_1})(\vec{a}_4(1) - \vec{z}_2^{k_2})}{(\vec{a}_4(1) - \vec{z}_2^{k_2})^2}$$

$$\mu_3 = (\mu_1 - \mu_1^{k_2})q_1q_2$$

$$q_3 = \frac{\left|\vec{z}_2(\mu_1) - \vec{z}_2^{k_2}\right| \cdot \left|\vec{a}_4(1) - \vec{z}_2^{k_2}\right|}{\vec{z}_1(\vec{z}_2(\mu_1) - \vec{z}_2^{k_2})}$$

$$q_4 = \frac{\vec{z}_1(\vec{z}_1^{k_1} - \vec{a}_1)}{\left|\vec{a}_2 - \vec{a}_1\right| \cdot \left|\vec{z}_1^{k_1} - \vec{a}_1\right|}$$

$$\lambda_3(\mu_1) = \frac{\mu_3}{\mu_1^{k_1} - \mu_1} \cdot q_3 \cdot q_4$$

$$cos\alpha_{2,3} = \frac{(\vec{z}_2(\mu_1) - \vec{z}_2^{k_2}) \cdot (\vec{a}_4(1) - \vec{z}_2^{k_2})}{\left|\vec{z}_2(\mu_1) - \vec{z}_2^{k_2}\right| \left|\vec{a}_4(1) - \vec{z}_2^{k_2}\right|}$$

$$A_3(\mu_1) = (\mu_1^{k_1} - \mu_1) \cdot \frac{|\vec{a}_2 - \vec{a}_1| \cdot |\vec{z}_1^{k_1} - \vec{a}_1|}{\vec{z}_1(\vec{z}_1^{k_1} - \vec{a}_1)} \quad (7)$$

$$\Omega_3(\lambda_3(\mu_1), \alpha_{2,3}) = \lambda_3(\mu_1) \cdot cos\alpha_{2,3}$$

$$\vec{Z}_3(\mu_1) = \vec{z}_1\{1 + A_3(\mu_1)[1 + \omega_2(\mu_1) + \Omega_3(\lambda_3, \alpha_{2,3})]\}$$

Give the approximate points $\vec{a}_1$, $\vec{a}_2$, $\vec{a}_3$, $\vec{a}_4(1)$ and also the values of parameters $\mu_1^{k_1} = \mu_1^*$ and $\mu_2^{k_2} = \mu_2^*$.

Introduce the denotation:

$\vec{a}_1 \to a1$; $\vec{a}_2 \to a2$; $\vec{a}_3 \to a3$; $(\vec{a}_4)_1 \to a4(1)$; $\mu_1 \to m1$;

$\mu_1^{k_1} \to m1k1$; $\mu_1^{k_2} \to m1k2$; $\mu_2 \to m2$; $\mu_2^{k_2} \to m2k2$;

$z_1^{k_1} \to z1k1$; $z_1^{k_2} \to z1k2$; $z_2^{k_2} \to z2k2$; $cos\alpha_{12} \to cosa12$;

$A(\mu_1^{k_2}) = A \to A$; $\lambda_2(\mu_1^{k_2}) = \lambda_2^{k_2} \to La2$;

$\omega_2(\lambda_2^{k_2}, \alpha_{12}) \to w2$;

$\vec{z}_2(\mu_1^{k_2}) \to z2$; $\vec{z}_1(\mu_1) \to z1$; $|\vec{z}_1(\mu_1)| \to z1M$;

$A(\mu_1) \to Am1$;

$\lambda_2(\ \mu_1) \to La2m1$; $\omega_2(\lambda_2(\mu_1), \alpha_{12}) \to w2m1$;

$\vec{z}_2(\mu_1) \to z2m1$; $|\vec{z}_2(\mu_1)| \to z2m1M$; $(\vec{a}_4)_2 \to a4(2)$;

$(\vec{a}_4)_3 \to a4(3)$; $(\vec{a}_4)_4 \to a4(4)$;......, $(\vec{a}_4)_\beta \to a4(\beta)$;

$q_1 = q1$; $q_2 = q2$ $\mu_3 \to m3$; $\lambda_3(\ \mu_1) \to La3m$; $q_3 = q3$;

$q_4 = q4$; $cos\alpha_{23} \to cosa23$;

$A_3(\mu_1) \to Am1p$; $\Omega_3(\lambda_3(\mu_1), \alpha_{23}) \to w3mp$;

$\vec{Z}_3(\mu_1) \to z3m1p$; $|\vec{Z}_3(\mu_1)| \to z3m1pM$;

$|\vec{z}_1(\mu_1)/\vec{Z}_3(\mu_1)| \to (z1M)/(z3m1pM) = B_1$;

$|\vec{z}_2(\mu_1)/\vec{Z}_3(\mu_1)| \to (z2m1M)/(z3m1pM) = B_2$;

$|\vec{z}_1(\mu_1)/\vec{z}_2(\mu_1)| \to (z1M)/(z2m1M) = B_3$;

$\vec{z}_1(1)/\vec{Z}_3(1) \to (z1(1))/(z3m1p(1)) = B_4$;

$\vec{z}_2(1)/\vec{Z}_3(1) \to (z2m1(1))/(z3m1p(1)) = B_5$;

$\vec{z}_1(1)/\vec{z}_2(1) \to (z1(1))/(z2m1(1)) = B_6$;

$\vec{z}_1(2)/\vec{Z}_3(2) \to (z1(2))/(z3m1p(2)) = B_7$;

$\vec{z}_2(2)/\vec{Z}_3(2) \to (z2m1(2))/(z3m1p(2)) = B_8$;

$\vec{z}_1(2)/\vec{z}_2(2) \to (z1(2))/(z2m1(2)) = B_9$;

$\vec{z}_1(3)/\vec{Z}_3(3) \to (z1(3))/(z3m1p(3)) = B_{10}$;

$\vec{z}_2(3)/\vec{Z}_3(3) \to (z2m1(3))/(z3m1p(3)) = B_{11}$;

$\vec{z}_1(3)/\vec{z}_2(3) \to (z1(3))/(z2m1(3)) = B_{12}$

$$\vec{z}_1(4)/\vec{Z}_3(4) \rightarrow (z1(4))/(z3m1p(4)) = B_{13};$$

$$\vec{z}_2(4)/\vec{Z}_3(4) \rightarrow (z2m1(4))/(z3m1p(4)) = B_{14}; \qquad (8)$$

$$\vec{z}_1(4)/\vec{z}_2(4) \rightarrow (z1(4))/(z2m1(4)) = B_{15}, \ldots\ldots$$

For the general *m*-dimensional case we write the symbolic representation of relations of the vectors in the following compact form:

$$\vec{z}_i(\beta)/\vec{z}_j(\beta) \rightarrow (zi(\beta))/(zjm1p(\beta)) = B_{ij}(\beta) \qquad (9)$$

Here the indices i and j $(i, j = 1,2,3)$ indicate the number of the vector $\vec{z}_i$ the index β the coordinate of the vector $\vec{z}_i$. And in *m*-dimensional case β takes integer values $\beta = 1,2,3,\ldots.m$.

Using the introduced denotation Eqs. (8) and (9), the algorithm and approximate numerical program for the system Eq. (7) in the *Matlab* language will be represented in the form:

a_1=[a_{11} a_{12} a_{13} $a_{14}\ldots a1\beta \ldots a1m$]

a_2=[a_{21} a_{22} a_{23} $a_{24}\ldots a2\beta \ldots a2m$]

a_3=[a_{31} a_{32} a_{33} $a_{34}\ldots a3\beta \ldots a3m$]

$m_1k_1=(m_1)^*$

$m_2k_2=(m_2)^*$

$a_4(1)=a_4(1)^*$

for $m_1=J_1:J_2:J_3$

$z_1k_1=a_1+m_1k_1*(a_2-a_1);$

$m_1k_2=m_1k_1+m_2k_2*((a_3-z_1k_1)*(a_3-z_1k_1)')/((a_3-z_1k_1)*(a_2-a_1)');$

$z_1k_2=a_1+m_1k_2*(a_2-a_1);$

$z_2k_2=z_1k_1+(m_1k_2-m_1k_1)*((a_3-z_1k_1)*(a_2-a_1)')/((a_3-z_1k_1)*(a_3-z_1k_1)')*(a_3-z_1k_1);$

$cosa_{12}=((z_1k_2-z_1k_1)*(z_2k_2-z_1k_1)')/(sqrt((z_1k_2-z_1k_1)*(z_1k_2-z_1k_1)')*sqrt((z_2k_2-z_1k_1)*(z_2k_2-z_1k_1)'));$

$A=(m_1k_1-m_1k_2)*(sqrt((a_2-a_1)*(a_2-a_1)')*sqrt((z_1k_1-a_1)*(z_1k_1-a_1)'))/(z_1k_2*(z_1k_1-a_1)');$

$p_1=m_2k_2/(m_1k_1-m_1k_2);$

$p_2=(sqrt((z_1k_2-z_1k_1)*(z_1k_2-z_1k_1)')*sqrt((a_3-z_1k_1)*(a_3-z_1k_1)'))/(z_1k_2*(z_1k_2-z_1k_1)');$

$p_3=(z_1k_2*(z_1k_1-a_1)')/(sqrt((a_2-a_1)*(a_2-a_1)')*sqrt((z_1k_1-a_1)*(z_1k_1-a_1)'));$

$La_2=p_1*p_2*p_3;$

$w_2=La_2*cosa_{12};$

$z_2=z_1k_2*(1+A*(1+w_2))$

$z_1=a_1+m_1*(a_2-a_1)$

$z_1\ M=sqrt((z_1)*(z_1)')$

$m_2=(m_1-m_1k_1)*(((a_3-z_1k_1)*(a_2-a_1)')/((a_3-z_1k_1)*(a_3-z_1k_1)'))$

$Am_1=(m_1k_1-m_1)*(sqrt((a_2-a_1)*(a_2-a_1)')*sqrt((z_1k_1-a_1)*(z_1k_1-a_1)'))/(z_1*(z_1k_1-a_1)');$

$p_1\ m_1=m_2/(m_1k_1-m_1);$

$p_2\ m_1=(sqrt((z_1-z_1k_1)*(z_1-z_1k_1)')*sqrt((a_3-z_1k_1)*(a_3-z_1k_1)'))/(z_1*(z_1-z_1k_1)');$

$p_3\ m_1=(z_1*(z_1k_1-a_1)')/(sqrt((a_2-a_1)*(a_2-a_1)')*sqrt((z_1k_1-a_1)*(z_1k_1-a_1)'));$

$La_2\ m_1=p_1\ m_1*p_2\ m_1*p_3\ m_1;$

$w_2\ m_1=La_2\ m_1*cosa_{12};$

$z_2\ m_1=z_1*(1+Am_1*(1+w_2\ m_1))$

$z_2\ m_1\ M=sqrt((z_2\ m_1)*(z_2\ m_1)')$

$a_4(2)=z_2k_2(2)+[(a_2(2)-a_1(2))/(a_2(1)-a_1(1))]*(a_4(1)-z_2k_2(1));$

$a_4(3)=z_2k_2(3)+[(a_2(3)-a_1(3))/(a_2(1)-a_1(1))]*(a_4(1)-z_2k_2(1));$

$a_4(4)=z_2k_2(4)+[(a_2(4)-a_1(4))/(a_2(1)-a_1(1))]*(a_4(1)-z_2k_2(1));$

$q_1=[(a_2-a_1)*(a_3-z_1k_1)']/[(a_3-z_1k_1)*(a_3-z_1k_1)'];$

$q_2=((a_3-z_1k_1)*(a_4-z_2k_2)')/((a_4-z_2k_2)*(a_4-z_2k_2)');$

$m_3=(m_1-m_1k_2)*q_1*q_2$

$q_3=(sqrt((z_2\ m_1-z_2k_2)*(z_2\ m_1-z_2k_2)')*sqrt((a_4-z_2k_2)*(a_4-z_2k_2)'))/(z_1*(z_2\ m_1-z_2k_2)');$

$q_4=[z_1*(z_1k_1-a_1)']/[sqrt((a_2-a_1)*(a_2-a_1)')*sqrt((z_1k_1-a_1)*(z_1k_1-a_1)')];$

$La_3\ m=[m_3/(m_1k_1-m_1)]*q_3*q_4;$

$cosa_{23}=((z_2\ m_1-z_2k_2)*(a_4-z_2k_2)')/[sqrt((z_2\ m_1-z_2k_2)*$
$(z_2\ m_1-z_2k_2)')*sqrt((a_4-z_2k_2)*(a_4-z_2k_2)')];$

$Am_1p=(m_1k_1-m_1)*(sqrt((a_2-a_1)*(a_2-a_1)')*sqrt((z_1k_1-a_1)*(z_1k_1-a_1)'))/$
$(z_1*(z_1k_1-a_1)');$

$w_3\ mp=La_3\ m*cosa_{23};$

$z_3\ m_1p=z_1*[1+Am_1p*(1+w_2\ m_1+w_3\ mp)]$

$z_3\ m_1pM=sqrt((z_3\ m_1p)*(z_3\ m_1p)')$

$B_1=(z_1\ M)/(z_3\ m_1pM)$

$B_2=(z_2\ m_1\ M/(z_3\ m_1pM))$

$B_3=(z_1\ M)/(z_2\ m_1\ M)$

$B_4=(z_1(1))/(z_3\ m_1p(1))$

$B_5=(z_2\ m_1(1))/(z_3\ m_1p(1))$

$B6=(z_1(1))/(z_2\ m_1(1))$

$B7=(z_1(2))/(z_3\ m_1p(2))$

$B8=(z_2\ m_1(2))/(z_3\ m_1p(2))$

$B9=(z_1(2))/(z_2\ m_1(2))$

$B_10=(z_1(3))/(z_3\ m_1p(3))$

$B_{11}=(z_2\ m_1(3))/(z_3\ m_1p(3))$

$B_{12}=(z_1(3))/(z_2\ m_1(3))$

$B_{13}=(z_1(4))/(z_3\ m_1p(4))$

$B_{14}=(z_2\ m_1(4))/(z_3\ m_1p(4))$

$B_{15}=(z_1(4))/(z_2\ m_1(4))$

$$B_{ij}(\beta)=(zi(\beta))/(zjm1\rho(\beta)) \tag{10}$$

Giving the statistical data of the vectors $\vec{a}_1$, $\vec{a}_2$, $\vec{a}_3$, $\vec{a}_4(1)$ and the parameters $\mu_1^{k_1}$ and $\mu_2^{k_2}$, by means of the above-suggested numerical program we can conduct wider investigations on multivatiant prediction of economic event at uncertainty conditions on the base of 2-component piecewise–linear model in m-dimensional vector space.

In the subsequent sections, we will consider concrete special variants and examples.

5.3 DEVELOPMENT OF SOFTWARE FOR COMPUTER MODELING AND MULTIVARIANT PREDICTION OF ECONOMIC EVENT AT UNCERTAINTY CONDITIONS ON THE BASE OF 2-COMPONENT PIECEWISE-LINEAR ECONOMIC-MATHEMATICAL MODELS IN 3-DIMENSIONAL VECTOR SPACE

5.3.1 ACTION ALGORITHM FOR COMPUTER MODELING BY CONSTRUCTING 2-COMPONENT PIECEWISE-LINEAR ECONOMIC-MATHEMATICAL MODELS IN 3-DIMENSIONAL VECTOR SPACE

On the base of Matlab we suggest an algorithm and numerical method of calculation for numerical construction of 2-component piecewise-linear economic-mathematical model with regard to unaccounted factors influence in 3-dimensional space of a vector function, and also to consider a concrete numerical example.

For the case of 2-component piecewise-linear vector-function in 3-dimensional vector space we write the main equations and mathematical expressions that are subjected to numerical programming [4, 9–11].

Let in 3-dimensional space R_3 a statistical table in the form of the points (vectors) $\{\vec{a}_n\}$ set describing some economic process be given. Let these points be represented in the form of the points set of adjacent 2-component piecewise-linear vector equation of the form [4, 9, 12]:

$$\vec{z}_1 = \vec{a}_1 + \mu_1(\vec{a}_2 - \vec{a}_1) \tag{11}$$

$$\vec{z}_2 = \vec{z}_1^{k_1} + \mu_2(\vec{a}_3 - \vec{z}_1^{k_1}) \tag{12}$$

where $\vec{z}_1 = \vec{z}_1(z_{11}, z_{12}, z_{13})$ and $\vec{z}_2 = \vec{z}_2(z_{21}, z_{22}, z_{23}$ are the equations of the first and second piecewise-linear straight lines in 3-dimensional vector space. The vectors $\vec{a}_1(a_{11}, a_{12}, a_{13})$, $\vec{a}_2 = \vec{a}_2(a_{21}, a_{22}, a_{23})$ and $\vec{a}_3 = \vec{a}_3(a_{31}, a_{32}, a_{33})$ are the given points (vectors) in 3-dimensional space of the form:

$$\vec{a}_1 = a_{11}\vec{i}_1 + a_{12}\vec{i}_2 + a_{13}\vec{i}_3$$

$$\vec{a}_2 = a_{21}\vec{i}_1 + a_{22}\vec{i}_2 + a_{23}\vec{i}_3 \tag{13}$$

$$\vec{a}_3 = a_{31}\vec{i}_1 + a_{32}\vec{i}_2 + a_{33}\vec{i}_3$$

Here, $\mu_1 \geqslant 0$ and $\mu_2 \geqslant 0$ are arbitrary parameters, $\vec{z}_1^{k_1}$ is the intersection point of the straight lines $\vec{z}_1$ and $\vec{z}_2$.

The goal of the investigation is the following. Giving the approximate point $\vec{a}_1$, $\vec{a}_2$, $\vec{a}_3$ and also the value of the parameters $\mu_1^{k_1} = \mu_1^*$ and $\mu_2^{k_2} = \mu_2^*$, develop a computer calculation algorithm for the following equations and mathematical expressions [4, 9, 10, 13]:

$$\vec{z}_1^{k_1} = \vec{a}_1 + \mu_1^{k_1}(\vec{a}_2 - \vec{a}_1)$$

$$\mu_1^{k_2} = \mu_1^{k_1} + \mu_2^{k_2}\frac{(\vec{a}_3 - z_1^{k_1})^2}{(\vec{a}_3 - z_1^{k_1})(\vec{a}_2 - \vec{a}_1)}$$

$$\vec{z}_1^{k_2} = \vec{a}_1 + \mu_1^{k_2}(\vec{a}_2 - \vec{a}_1)$$

$$\vec{z}_2^{k_2} = \vec{z}_1^{k_1} + (\mu_1^{k_2} - \mu_1^{k_1})\frac{(\vec{a}_3 - \vec{z}_1^{k_1})(\vec{a}_3 - a_1)}{(\vec{a}_3 - \vec{z}_1^{k_1})^2}(\vec{a}_3 - \vec{z}_1^{k_1})$$

$$cos\alpha_{1,2} = \frac{(\vec{z}_1^{k_2} - \vec{z}_1^{k_1})(\vec{z}_1^{k_2} - \vec{z}_1^{k_1})}{\left|\vec{z}_1^{k_2} - \vec{z}_1^{k_1}\right|\left|\vec{z}_1^{k_2} - \vec{z}_1^{k_1}\right|}$$

$$A = (\mu_1^{k_1} - \mu_1^{k_2})\frac{\left|\vec{a}_2 - \vec{a}_1\right|\left|\vec{z}_1^{k_1} - \vec{a}_1\right|}{\vec{z}_1^{k_2}(\vec{z}_1^{k_1} - \vec{a}_1)}$$

$$\lambda_2 = \frac{\mu_2^{k_2}}{\mu_1^{k_1} - \mu_2^{k_2}} \cdot \frac{\left|\vec{z}_1^{k_2} - \vec{z}_1^{k_1}\right|\left|\vec{a}_3 - \vec{z}_1^{k_1}\right|}{\vec{z}_1^{k_2}(\vec{z}_1^{k_2} - \vec{z}_1^{k_1})} \cdot \frac{\vec{z}_1^{k_2}(\vec{z}_1^{k_1} - \vec{a}_1)}{\left|\vec{a}_2 - \vec{a}_1\right|\left|\vec{z}_1^{k_1} - \vec{a}_1\right|}$$

$$\omega_2(\lambda_2, \alpha_{1,2}) = \lambda_2 cos\alpha_{1,2}$$

$$\vec{z}_2 = \vec{z}_1^{k_2}\{1 + A[1 + \omega_2(\lambda_2, \alpha_{1,2})]\}$$

$$\vec{z}_1 = \vec{a}_1 + \mu_1(\vec{a}_2 - \vec{a}_1)$$

$$\mu_2 = (\mu_1 - \mu_1^{k_1}) + \frac{(\vec{a}_3 - \vec{z}_1^{k_1})(\vec{a}_2 - \vec{a}_1)}{(\vec{a}_3 - \vec{z}_1^{k_1})^2}$$

$$A(\mu_1) = (\mu_1^{k_1} - \mu_1)\frac{\left|\vec{a}_2 - \vec{a}_1\right|\left|\vec{z}_1^{k_1} - \vec{a}_1\right|}{\vec{z}_1(\vec{z}_1^{k_1} - \vec{a}_1)}$$

$$\lambda_2(\mu_1) = \frac{\mu_2}{\mu_1^{k_1} - \mu_1} \cdot \frac{\left|\vec{z}_1 - \vec{z}_1^{k_1}\right|\left|\vec{a}_3 - \vec{z}_1^{k_1}\right|}{\vec{z}_1(\vec{z}_1 - \vec{z}_1^{k_1})} \cdot \frac{\vec{z}_1(\vec{z}_1^{k_1} - \vec{a}_1)}{\left|\vec{a}_2 - \vec{a}_1\right|\left|\vec{z}_1^{k_1} - \vec{a}_1\right|}$$

$$\omega(\mu_1) = \lambda_2(\mu_1) cos\alpha_1$$

$$\vec{z}_2(\mu_1) = \vec{z}_1\{1 + A\ (\mu_1)[1 + \omega_2(\mu_1)]\} \tag{14}$$

Introduce the following denotation:

$$\begin{aligned}
&\mu_1^{k_2} \rightarrow m1k2;\ \vec{z}_1^{k_1} \rightarrow z1k1;\ \vec{z}_1^{k_2} \rightarrow z1k2;\ \vec{z}_2^{k_2} \rightarrow z2k2;\\
&\vec{z}_1 \rightarrow z1;\ \mu_2 \rightarrow m2;\ \mu_2^{k_2} \rightarrow m2k2;\ A(\mu_1) \rightarrow Am1;\\
&\lambda_2 \rightarrow La2;\ \lambda_2(\mu_1) \rightarrow La2m1;\\
&\omega_2(\lambda_2, \alpha_{12}) = \omega_2(\mu_1) \rightarrow w2m1;\\
&\vec{z}_2 \rightarrow z2;\ \vec{z}_2(\mu_1) \rightarrow z2m1
\end{aligned} \tag{15}$$

Using the introduced denotation Eq. (15), the appropriate computer algorithm for the system Eq. (14) for numerical construction of 2-component piecewise-linear economic-mathematical models with regard to unaccounted factors influence in 3-dimensional vector space, in the *Matlab* program will look like:

$a_1=[a_{11}\ a_{12}\ a_{13}]$

$a_2=[a_{21}\ a_{22}\ a_{23}]$

$a_3=[a_{31}\ a_{32}\ a_{33}]$

$m_1k_1=(m_1)^*$

$m_2k_2=(m_2)^*$

for $m_1=J_1: j_2:J_3$

$z_1k_1=a_1+m_1k_1*(a_2-a_1);$

$m_1k_2=m_1k_1+m_2k_2*((a_3-z_1k_1)*(a_3-z_1k_1)')/((a_3-z_1k_1)*(a_2-a_1)');$

$z_1k_2=a_1+m_1k_2*(a_2-a_1);$

$z_2k_2=z_1k_1+(m_1k_2-m_1k_1)*((a_3-z_1k_1)*(a_2-a_1)')/((a_3-z_1k_1)*(a_3-z_1k_1)')*(a_3-z_1k_1);$

$cosa_{12}=((z_1k_2-z_1k_1)*(z_2k_2-z_1k_1)')/(sqrt((z_1k_2-z_1k_1)*(z_1k_2-z_1k_1)')*sqrt((z_2k_2-z_1k_1)*(z_2k_2-z_1k_1)'))$

$A=(m_1k_1-m_1k_2)*(sqrt((a_2-a_1)*(a_2-a_1)')*sqrt((z_1k_1-a_1)*(z_1k_1-a_1)'))/(z_1k_2*(z_1k_1-a_1)');$

$p_1=m_2k_2/(m_1k_1-m_1k_2);$

$p_2=(sqrt((z_1k_2-z_1k_1)*(z_1k_2-z_1k_1)')*sqrt((a_3-z_1k_1)*(a_3-z_1k_1)'))/(z_1k_2*(z_1k_2-z_1k_1)');$

$p_3=(z_1k_2*(z_1k_1-a_1)')/(sqrt((a_2-a_1)*(a_2-a_1)')*sqrt((z_1k_1-a_1)*(z_1k_1-a_1)'));$

$La_2=p_1*p_2*p_3;$

$w_2=La_2*cosa_{12};$

$z_2=z_1k_2*(1+A*(1+w_2));$

$z_1=a_1+m_1*(a_2-a_1)$

$m_2=(m_1-m_1k_1)*(((a_3-z_1k_1)*(a_2-a_1)')/((a_3-z_1k_1)*$

$(a_3-z_1k_1)'))$

$Am_1=(m_1k_1-m_1)*(sqrt((a_2-a_1)*(a_2-a_1)')*sqrt((z_1k_1-a_1)*(z_1k_1-a_1)'))/(z_1*(z_1k_1-a_1)')$

$p_1\ m_1=m_2/(m_1k_1-m_1);$

$p_2\ m_1=(sqrt((z_1-z_1k_1)*(z_1-z_1k_1)')*sqrt((a_3-z_1k_1)*(a_3-z_1k_1)'))/(z_1*(z_1-z_1k_1)');$

$p_3\ m_1=(z_1*(z_1k_1-a_1)')/(sqrt((a_2-a_1)*(a_{2-1})')*sqrt((z_1k_1-a_1)*(z_1k_1-a_1)'));$

$La_2\, m_1 = p_1\, m_1 * p_2\, m_1 * p_3\, m_1;$

$w_2\, m_1 = La_2\, m_1 * cosa_{12}$

$z_2\, m_1 = z_1 * (1 + Am_1 * (1 + w_2\, m_1))$

end (5.3.6)

3. **Example:** As an example we give the following table of statistical data. Let the vectors $\vec{a}_1$, $\vec{a}_2$, $\vec{a}_3$ and the parameters $\mu_1^{k_1}$ and $\mu_2^{k_2}$ have the following numerical values:

$a_1 = [1 -1\ 1];$

$a_2 = [3 -2\ 4,\ 5];$

$a_3 = [6 -4\ 7];$

$m_1 k_1 = 1.5$

$m_2 k_2 = 2$

for $m_1 = 1,5{:}0,5{:}8$.

The task of the investigation is to represent the points of the second piecewise-linear straight line depending on the first piecewise-linear vector-function $\vec{z}_1(\mu_1)$ and unaccounted factors influence function $\omega_2(\lambda_2, \alpha_{1,2})$ for arbitrary values of the parameter μ_1 changing in the interval $\mu_1^{k_1} = 1,5 \le \mu_1 \le \mu_1^* = 8$, in the form:

$$\vec{z}_2(\mu_1) = \vec{z}_1(\mu_1)\{1 + A(\mu_1)[1 + \omega_2(\mu_1)]\}$$

Applying the above-stated numerical program to the given problem, we numerically define the points of the second piecewise-linear straight line $\vec{z}_2(\mu_1)$ depending on the parameter $\mu_1 \ge 1,5$, that are represented in the form of Table 1.

It should be noted that the numerically constructed vectors $\vec{z}_2(\mu_1)$ for different values of the parameter $1,5 \le \mu_1 \ge 8$ completely coincide with numerical results obtained earlier in the Refs. [4, 5, 9, 14] by developing theory of construction of piecewise-linear economic-mathematical mod-

Table 5.1 Numerical values of the points (vectors) of the second piecewise-linear straight line $\Xi_2(\mu_1)$ depending on the parameter $\mu_1 \geq 1{,}5$ in 3-dimensional vector space, defined by the formula: $\Xi_2(\mu_1) = \Xi_1(\mu_1)\{1 + A(\mu_1)[1 + \omega_2(\mu_1)]\}$, calculated for the following given values of the vectors $\vec{a}_1 = \vec{i}_1 + \vec{i}_2 + \vec{i}_3$, $\vec{a}_2 = 3\vec{i}_1 + 2\vec{i}_2 + 4{,}5\vec{i}_3$, $\vec{a}_3 = 6\vec{i}_1 + 4\vec{i}_2 + 7\vec{i}_3$, and parameters: $\mu_1^{k_1} = 1{,}5$ и $\mu_2^{k_2} = 2$

N	μ_1	μ_2	$A(\mu_1)$	$\omega_2(\mu_1)$	$\Xi_1(\mu_1) = \vec{a}_1 + \mu_1(\vec{a}_2 - \vec{a}_1)$	$\Xi_2(\mu_1) = \Xi_1(\mu_1)\{1 + A(\mu_1)[1 + \omega_2(\mu_1)]\}$
1	1,5	0	0	0	Ξ_1(1,5) =[4 2,5 6,25]	Ξ_2(1,5)=[4 2,5 6,25]
2	2	0,5963	-0,2104	-0,5618	Ξ_1(2)=[5 3 8]	Ξ_2(2)=[4,539 2,7234 7,2625]
3	2,5	1,1927	-0,3476	-0,5618	Ξ_1(2,5)=[6 3,5 9,75]	Ξ_2(2,5)=[5,086 2,9668 8,2647]
4	3	1,789	-0,4442	-0,5618	Ξ_1(3) =[7 4 11,5]	Ξ_2(3) =[5,6372 3,2213 9,2613]
5	3,1769	2	-0,4719	-0,5618	Ξ_1(3,1769) =[7,3538 4,1769 12,1192]	Ξ_2(3,1769) =[5,8331 3,3132 9,613]
6	3,5	2,3853	-0,5159	-0,5618	Ξ_1(3,5) =[8 4,5 13,25]	Ξ_2(3,5) =[6,1913 3,4826 10,2544]
7	4	2,9817	-0,5712	-0,5618	Ξ_1(4)=[9 5 15]	Ξ_2(4)=[6,7471 3,7484 11,2452]
8	4,5	3,578	-0,6152	-0,5618	Ξ_1(4,5)=[10 5,5 16,75]	Ξ_2(4,5)=[7,3041 4,0173 12,2344]

...continued

9	5	4,1743	-0,6509	-0,5618	$\Xi_1(5)=[11\ \ 6\ \ 18{,}5]$	$\Xi_2(5)=[7{,}862\ \ 4{,}2884\ \ 13{,}2225]$
10	5,5	4,7706	-0,6806	-0,5618	$\Xi_1(5{,}5)=[12\ \ 6{,}5\ \ 20{,}25]$	$\Xi_2(5{,}5)=[8{,}4206\ \ 4{,}5612\ \ 14{,}2098]$
11	6	5,3670	-0,7057	-0,5618	$\Xi_1(6)=[13\ \ 7\ \ 22]$	$\Xi_2(6)=[8{,}9796\ \ 4{,}8352\ \ 15{,}1963]$
12	6,5	5,9633	-0,7271	-0,5618	$\Xi_1(6{,}5)=[14\ \ 7{,}5\ \ 23{,}75]$	$\Xi_2(6{,}5)=[9{,}5391\ \ 5{,}1102\ \ 16{,}1824]$
13	7	6,5596	-0,7456	-0,5618	$\Xi_1(7)=[15\ \ 8\ \ 25{,}5]$	$\Xi_2(7)=[10{,}0989\ \ 5{,}3861\ \ 17{,}1681]$
14	7,5	7,1560	-0,7617	-0,5618	$\Xi_1(7{,}5)=[16\ \ 8{,}5\ \ 27{,}25]$	$\Xi_2(7{,}5)=[10{,}6589\ \ 5{,}6625\ \ 18{,}1534]$
15	8	7,7523	-0,7760	-0,5618	$\Xi_1(8)=[17\ \ 9\ \ 29]$	$\Xi_2(8)=[11{,}2191\ \ 5{,}9395\ \ 19{,}1385]$

els at uncertainty conditions in finite dimensional vector space stated in Chapter 3.

5.3.2 ALGORITHM OF MULTIVARIANT COMPUTER MODELING OF PREDICTION VARIABLES OF ECONOMIC EVENT ON THE BASE OF 2-COMPONENT PIECEWISE-LINEAR ECONOMIC-MATHEMATICAL MODELS IN 3-DIMENSIONAL VECTOR SPACE

In this section, based around the *Matlab* program, we suggest an algorithm and numerical programe for multivariant prediction of economic event at uncertainty conditions on the vase of 2-component piecewise-linear model in 3-dimenaionl vector space [10, 15].

According to the theory, Refs. [10, 15] for the case of a 2-component piecewise-linear vector-function at uncertainty conditions in 3-dimensional vector space for a multivariant prediction of economic event we have the following equations and expressions:

$$\vec{z}_1^{k_1} = \vec{a}_1 + \mu_1^{k_1} (\vec{a}_2 - \vec{a}_1)$$

$$\mu_1^{k_2} = \mu_1^{k_1} + \mu_2^{k_2} \frac{(\vec{a}_3 - \vec{z}_1^{k_1})^2}{(\vec{a}_3 - \vec{z}_1^{k_1})(\vec{a}_2 - \vec{a}_1)}$$

$$\vec{z}_1^{k_2} = \vec{a}_1 + \mu_1^{k_2} (\vec{a}_2 - \vec{a}_1)$$

$$\vec{z}_2^{k_2} = \vec{z}_1^{k_1} + (\mu_1^{k_2} - \mu_1^{k_1}) \frac{(\vec{a}_3 - \vec{z}_1^{k_1})(\vec{a}_2 - \vec{a}_1)}{(\vec{a}_3 - \vec{z}_1^{k_1})^2} (\vec{a}_3 - \vec{z}_1^{k_1})$$

$$cos\alpha_{1,2} = \frac{(\vec{z}_1^{k_2} - \vec{z}_1^{k_1})(\vec{z}_2^{k_2} - \vec{z}_1^{k_1})}{\left|\vec{z}_1^{k_2} - \vec{z}_1^{k_1}\right| \cdot \left|\vec{z}_2^{k_2} - \vec{z}_1^{k_1}\right|}$$

$$A(\mu_1^{k_2}) = A = (\mu_1^{k_1} - \mu_1^{k_2})\frac{|\vec{a}_2 - \vec{a}_1||\vec{z}_1^{k_1} - \vec{a}_1|}{\vec{z}_1^{k_2}(\vec{z}_1^{k_1} - \vec{a}_1)}$$

$$\lambda_2(\mu_1^{k_2}) = \lambda_2^{k_2} = \frac{\mu_2^{k_2}}{\mu_1^{k_1} - \mu_1^{k_2}} \cdot \frac{|\vec{z}_1^{k_2} - \vec{z}_1^{k_1}||\vec{a}_3 - \vec{z}_1^{k_1}|}{\vec{z}_1^{k_2}(\vec{z}_1^{k_2} - \vec{z}_1^{k_1})} \cdot \frac{\vec{z}_1^{k_2}(\vec{z}_1^{k_1} - \vec{a}_1)}{|\vec{a}_2 - \vec{a}_1||\vec{z}_1^{k_1} - \vec{a}_1|}$$

$$\omega_2(\lambda_2^{k_2}, \alpha_{1,2}) = \lambda_2^{k_2} cos\alpha_{1,2}$$

$$\vec{z}_2(\mu_1^{k_2}) = \vec{z}_2 = \vec{z}_1^{k_2}\{1 + A\,[1 + \omega_2(\lambda_2^{k_2}, \alpha_{1,2})]\}$$

$$\vec{z}_1(\mu_1) = \vec{z}_1 = \vec{a}_1 + \mu_1(\vec{a}_2 - \vec{a}_1)$$

$$\mu_2 = (\mu_1 - \mu_1^{k_1}) \cdot \frac{(\vec{a}_3 - \vec{z}_1^{k_1})(\vec{a}_2 - \vec{a}_1)}{(\vec{a}_3 - \vec{z}_1^{k_1})^2}$$

$$A(\mu_1) = (\mu_1^{k_1} - \mu_1)\frac{|\vec{a}_2 - \vec{a}_1||\vec{z}_1^{k_1} - \vec{a}_1|}{\vec{z}_1(\vec{z}_1^{k_1} - \vec{a}_1)}$$

$$\lambda_2(\mu_1) = \frac{\mu_2}{\mu_1^{k_1} - \mu_1} \cdot \frac{|\vec{z}_1 - \vec{z}_1^{k_1}||\vec{a}_3 - \vec{z}_1^{k_1}|}{\vec{z}_1(\vec{z}_1 - \vec{z}_1^{k_1})} \cdot \frac{\vec{z}_1(\vec{z}_1^{k_1} - \vec{a}_1)}{|\vec{a}_2 - \vec{a}_1||\vec{z}_1^{k_1} - \vec{a}_1|}$$

$$\omega_2(\lambda_2(\mu_1), \alpha_{12}) = \omega_2(\mu_1) = \lambda_2(\mu_1) cos\alpha_{1,2}$$

$$\vec{z}_2(\mu_1) = \vec{z}_1\{1 + A\,(\mu_1)[1 + \omega_2(\mu_1)]\}$$

$$a_{42}(1) = (\vec{a}_4)_2 = z_{22}^{k_2} + \frac{a_{22} - a_{12}}{a_{21} - a_{11}}(a_{41}(1) - z_{21}^{k_2})$$

$$a_{43}(1) = (\vec{a}_4)_3 = z_{23}^{k_2} + \frac{a_{23} - a_{13}}{a_{21} - a_{11}}(a_{41}(1) - z_{21}^{k_2})$$

$$q_1 = \frac{(\vec{a}_2 - \vec{a}_1)(\vec{a}_3 - \vec{z}_1^{k_1})}{(\vec{a}_3 - \vec{z}_1^{k_1})^2}$$

$$q_2 = \frac{(\vec{a}_3 - \vec{z}_1^{k_1})(\vec{a}_4(1) - \vec{z}_2^{k_2})}{(\vec{a}_4(1) - \vec{z}_2^{k_2})^2}$$

$$\mu_3 = (\mu_1 - \mu_1^{k_2})q_1 q_2$$

$$q_3 = \frac{\left|\vec{z}_2(\mu_1) - \vec{z}_2^{k_2}\right| \cdot \left|\vec{a}_4(1) - \vec{z}_2^{k_2}\right|}{\vec{z}_1(\vec{z}_2(\mu_1) - \vec{z}_2^{k_2})}$$

$$q_4 = \frac{\vec{z}_1(\vec{z}_1^{k_1} - \vec{a}_1)}{\left|\vec{a}_2 - \vec{a}_1\right| \cdot \left|\vec{z}_1^{k_1} - \vec{a}_1\right|}$$

$$\lambda_3(\mu_1) = \frac{\mu_3}{\mu_1^{k_1} - \mu_1} \cdot q_3 \cdot q_4$$

$$cos\alpha_{2,3} = \frac{(\vec{z}_2(\mu_1) - \vec{z}_2^{k_2}) \cdot (\vec{a}_4(1) - \vec{z}_2^{k_2})}{\left|\vec{z}_2(\mu_1) - \vec{z}_2^{k_2}\right| \left|\vec{a}_4(1) - \vec{z}_2^{k_2}\right|}$$

$$A_3(\mu_1) = (\mu_1^{k_1} - \mu_1) \cdot \frac{\left|\vec{a}_2 - \vec{a}_1\right| \cdot \left|\vec{z}_1^{k_1} - \vec{a}_1\right|}{\vec{z}_1(\vec{z}_1^{k_1} - \vec{a}_1)}$$

$$\Omega_3(\lambda_3(\mu_1),\alpha_{2,3}) = \lambda_3(\mu_1)\cdot cos\alpha_{2,3}$$

$$\vec{Z}_3(\mu_1) = \vec{z}_1\{1 + A_3(\mu_1)[1+\omega_2(\mu_1)+\Omega_3(\lambda_3,\alpha_{2,3})]\} \tag{17}$$

Give the approximation points $\vec{a}_1$, $\vec{a}_2$, $\vec{a}_3$, $\vec{a}_4(1)$ and also the values of the parameters $\mu_1^{k_1} = \mu_1^*$ and $\mu_2^{k_2} = \mu_2^*$. Here, $\vec{a}_4(1)$ is the given value of one of the coordinates of the vector $\vec{a}_4$. For that we introduce the following denotation:

$\vec{a}_1 \to a1$; $\vec{a}_2 \to a2$; $\vec{a}_3 \to a3$; $(\vec{a}_4)_1 \to a4(1)$;

$\mu_1 \to m1$; $\mu_1^{k_1} \to m1k1$; $\mu_1^{k_2} \to m1k2$; $\mu_2 \to m2$;

$\mu_2^{k_2} \to m2k2$; $z_1^{k_1} \to z1k1$; $z_1^{k_2} \to z1k2$; $z_2^{k_2} \to z2k2$;

$cos\alpha_{12} \to cosa12$; $A(\mu_1^{k_2}) = A \to A$; $\lambda_2(\mu_1^{k_2}) = \lambda_2^{k_2} \to La2$;

$\omega_2(\lambda_2^{k_2},\alpha_{12}) \to w2$; $\vec{z}_2(\mu_1^{k_2}) \to z2$; $\vec{z}_1(\mu_1) \to z1$;

$|\vec{z}_1(\mu_1)| \to z1M$; $A(\mu_1) \to Am1$; $\lambda_2(\mu_1) \to La2m1$;

$\omega_2(\lambda_2(\mu_1),\alpha_{12}) \to w2m1$; $\vec{z}_2(\mu_1) \to z2m1$;

$|\vec{z}_2(\mu_1)| \to z2m1M$; $(\vec{a}_4)_2 \to a4(2)$; $(\vec{a}_4)_3 \to a4(3)$;

$q_1 = q1$; $q_2 = q2$; $\mu_3 \to m3$; $\lambda_3(\mu_1) \to La3m$; $q_3 = q3$;

$q_4 = q4$; $cos\alpha_{23} \to cosa23$; $A_3(\mu_1) \to Am1p$;

$\Omega_3(\lambda_3(\mu_1),\alpha_{23}) \to w3mp$; $\vec{Z}_3(\mu_1) \to z3m1p$;

$|\vec{Z}_3(\mu_1)| \to z3m1pM$;

$|\vec{z}_1(\mu_1)/\vec{Z}_3(\mu_1)| \to (z1M)/(z3m1pM) = B_1$;

$$
\begin{aligned}
&\left|\vec{z}_2(\mu_1)/\vec{Z}_3(\mu_1)\right| \to (z2m1M)/(z3m1pM) = B_2;\\
&\left|\vec{z}_1(\mu_1)/\vec{z}_2(\mu_1)\right| \to (z1M)/(z2m1M) = B_3;\\
&\vec{z}_1(1)/\vec{Z}_3(1) \to (z1(1))/(z3m1p(1)) = B_4;\\
&\vec{z}_2(1)/\vec{Z}_3(1) \to (z2m1(1))/(z3m1p(1)) = B_5;\\
&\vec{z}_1(1)/\vec{z}_2(1) \to (z1(1))/(z2m1(1)) = B_6;\\
&\vec{z}_1(2)/\vec{Z}_3(2) \to (z1(2))/(z3m1p(2)) = B_7;\\
&\vec{z}_2(2)/\vec{Z}_3(2) \to (z2m1(2))/(z3m1p(2)) = B_8;\\
&\vec{z}_1(2)/\vec{z}_2(2) \to (z1(2))/(z2m1(2)) = B_9;\\
&\vec{z}_1(3)/\vec{Z}_3(3) \to (z1(3))/(z3m1p(3)) = B_{10};\\
&\vec{z}_2(3)/\vec{Z}_3(3) \to (z2m1(3))/(z3m1p(3)) = B_{11};\\
&\vec{z}_1(3)/\vec{z}_2(3) \to (z1(3))/(z2m1(3)) = B_{12}
\end{aligned} \tag{18}
$$

Using the introduced denotation, an algorithm and appropriate numerical program for the system Eq. (18) in the *Matlab* language will be represented in the form [4, 9, 10, 16]:

$a_1=[a_{11}\ a_{12}\ a_{13}]$

$a_2=[a_{21}\ a_{22}\ a_{23}]$

$a_3=[a_{31}\ a_{32}\ a_{33}]$

$m_1k_1=(m_1)*$

$m_2k_2=(m_2)*$

$a_4(1)=a_4(1)*$

for $m_1=J_1:J_2:J_3$

$z_1k_1=a_1+m_1k_1*(a_2-a_1);$

$m_1k_2=m_1k_1+m_2k_2*((a_3-z_1k_1)*(a_3-z_1k_1)')/((a_3-z_1k_1)*(a_2-a_1)');$

$z_1k_2=a_1+m_1k_2*(a_2-a_1);$

$z_2k_2=z_1k_1+(m_1k_2-m_1k_1)*((a_3-z_1k_1)*(a_2-a_1)')/((a_3-z_1k_1)*(a_3-z_1k_1)')*(a_3-z_1k_1);$

$cosa_{12}=((z_1k_2-z_1k_1)*(z_2k_2-z_1k_1)')/(sqrt((z_1k_2-z_1k_1)*(z_1k_2-z_1k_1)')*sqrt((z_2k_2-z_1k_1)*(z_2k_2-z_1k_1)'));$

$A=(m_1k_1-m_1k_2)*(sqrt((a_2-a_1)*(a_2-a_1)')*sqrt((z_1k_1-a_1)*(z_1k_1-a_1)'))/(z_1k_2*(z_1k_1-a_1)');$

$p_1=m_2k_2/(m_1k_1-m_1k_2);$

$p_2=(sqrt((z_1k_2-z_1k_1)*(z_1k_2-z_1k_1)')*sqrt((a_3-z_1k_1)*(a_3-z_1k_1)'))/(z_1k_2*(z_1k_2-z_1k_1)');$

$p_3=(z_1k_2*(z_1k_1-a_1)')/(sqrt((a_2-a_1)*(a_2-a_1)')*sqrt((z_1k_1-a_1)*(z_1k_1-a_1)'));$

$La_2=p_1*p_2*p_3;$

$w_2=La_2*cosa_{12};$

$z_2=z_1k_2*(_1+A*(_1+w_2))$

$z_1=a_1+m_1*(a_2-a_1)$

$z_1\ M=sqrt((z_1)*(z_1)')$

$m_2=(m_1-m_1k_1)*(((a_3-z_1k_1)*(a_2-a_1)')/((a_3-z_1k_1)*(a_3-z_1k_1)'))$

$Am_1=(m_1k_1-m_1)*(sqrt((a_2-a_1)*(a_2-a_1)')*sqrt((z_1k_1-a_1)*(z_1k_1-a_1)'))/(z_1*(z_1k_1-a_1)');$

$p_1\ m_1=m_2/(m_1k_1-m_1);$

$p_2m_1=(sqrt((z_1-z_1k_1)*(z_1-z_1k_1)')*sqrt((a_3-z_1k_1)*(a_3-z_1k_1)'))/(z_1*(z_1-z_1k_1)');$

$p_3m_1=(z_1*(z_1k_1-a_1)')/(sqrt((a_2-a_1)*(a_2-a_1)')*sqrt((z_1k_1-a_1)*(z_1k_1-a_1)'));$

$La_2\ m_1=p_1\ m_1*p_2\ m_1*p_3\ m_1;$

$w_2\ m_1=La_2\ m_1*cosa_{12};$

$z_2\ m_1=z_1*(_1+Am_1*(_1+w_2\ m_1))$

$z_2\ m_1\ M=sqrt((z_2\ m_1)*(z_2\ m_1)')$

$a_4(2)=z_2k_2(2)+[(a_2(2)-a_1(2))/(a_2(1)-a_1(1))]*(a_4(1)-z_2k_2(1));$

$a_4(3)=z_2k_2(3)+[(a_2(3)-a_1(3))/(a_2(1)-a_1(1))]*(a_4(1)-z_2k_2(1));$

$q_1=[(a_2-a_1)*(a_3-z_1k_1)']/[(a_3-z_1k_1)*(a_3-z_1k_1)'];$

$q_2=((a_3-z_1k_1)*(a_4-z_2k_2)')/((a_4-z_2k_2)*(a_4-z_2k_2)');$

$m_3=(m_1-m_1k_2)*q_1*q_2$

$q_3=(sqrt((z_2m_1-z_2k_2)*(z_2m_1-z_2k_2)')*sqrt((a_4-z_2k_2)*(a_4-z_2k_2)'))/(z_1*(z_2\ m_1-z_2k_2)');$

$q_4=[z_1*(z_1k_1-a_1)']/[sqrt((a_2-a_1)*(a_2-a_1)')*sqrt((z_1k_1-a_1)*(z_1k_1-a_1)')];$

$La_3\ m=[m_3/(m_1k_1-m_1)]*q_3*q_4;$

$cosa_{23}=((z_2\ m_1-z_2k_2)*(a_4-z_2k_2)')/[sqrt((z_2\ m_1-z_2k_2)*(z_2\ m_1-z_2k_2)')*sqrt((a_4-z_2k_2)*(a_4-z_2k_2)')];$

$Am_1p=(m_1k_1-m_1)*(sqrt((a_2-a_1)*(a_2-a_1)')*sqrt((z_1k_1-a_1)*(z_1k_1-a_1)'))/(z_1*(z_1k_1-a_1)');$

$w_3\ mp=La_3\ m*cosa_{23};$

$z_3\ m_1p=z_1*[_1+Am_1p*(_1+w_2\ m_1+w_3\ mp)]$

$z_3\ m_1pM=sqrt((z_3\ m_1p)*(z_3\ m_1p)')$

$B_1=(z_1\ M)/(z_3\ m_1pM)$

$B_2=(z_2\ m_1\ M/(z_3\ m_1pM))$

$B_3=(z_1\ M)/(z_2\ m_1\ M)$

$B_4=(z_1(1))/(z_3\ m_1p(1))$

$B_5=(z_2\ m_1(1))/(z_3\ m_1p(1))$

B_6=(z_1(1))/(z_2 m_1(1))

B_7=(z_1(2))/(z_3 m_1p(2))

B_8=(z_2 m_1(2))/(z_3 m_1p(2))

B_9=(z_1(2))/(z_2 m_1(2))

B_{10}=(z_1(3))/(z_3 m_1p(3))

B_{11}=(z_2 m_1(3))/(z_3 m_1p(3))

B_{12}=(z_1(3))/(z_2 m_1(3))

end. (19)

Giving the statistical data of the vectors $\vec{a}_1$, $\vec{a}_2$, $\vec{a}_3$, $\vec{a}_4(1)$ and the parameters $\mu_1^{k_1}$ and $\mu_2^{k_2}$ by means of the above suggested numerical program we can conduct wider investigations on multivariant prediction of economic event at uncertainty condition on the base of 2-component piecewise-linear model in 3-dimensional vector space.

Example. As an example consider the case with the following given statistical vectors $\vec{a}_1$, $\vec{a}_2$, $\vec{a}_3$, $\vec{a}_4(1)$ and parameters $\mu_1^{k_1}$ and $\mu_2^{k_2}$:

a_1=[1–1 1]

a_2=[3–2 4.5]

a_3=[6–4 7]

m_1k_1=1.5

m_2k_2=2

a_4(1)=10

for m_1=1,5:0,5:8.

Conduct the following numerical calculation on establishment of all possible variants of prediction numerical values of economic event on the subsequent stage represented in Tables 2 and 3.

TABLE 2 Numerical values of modulus and appropriate coordinates of predictable points-vectors for different values of parameters $3{,}1769, \leq \mu_1 \leq 8$ and $0 \leq \mu_3 \leq 2{,}7094$.

N	**Numerical values of the vectors $\vec{z}_1$, $\vec{z}_2$, $\vec{z}_3$ and their module**	μ_1	μ_2	μ_3
1	$\vec{z}_1(\mu_1)$=[7.3538–4.1769–12.1769] $\vec{z}_2(\mu_1)$=[5.8331–3.3132–9.6130] $\vec{Z}_3(\mu_1)$=[5.8331–3.3132–9.6190] $\lvert\vec{z}_1(\mu_1)\rvert$=14.8257 $\lvert\vec{z}_2(\mu_1)\rvert$=11.7223 $\lvert\vec{Z}_3(\mu_1)\rvert$=11.7227	3.1769	2	0
2	$\vec{z}_1(\mu_1)$=[8.–4.5–13.25] $\vec{z}_2(\mu_1)$=[6.1913–3.4826–10.2544] $\vec{Z}_3(\mu_1)$=[6.6631–3.7480–11.0357] $\lvert\vec{z}_1(\mu_1)\rvert$=16.1187 $\lvert\vec{z}_2(\mu_1)\rvert$=12.4745 $\lvert\vec{Z}_3(\mu_1)\rvert$=13.4250	3.5	2.3853	0.1815
3	$\vec{z}_1(\mu_1)$=[9–5 15] $\vec{z}_2(\mu_1)$=[6.7471–3.7484–11.2452] $\vec{Z}_3(\mu_1)$=[7.7717–4.3176–12.9528] $\lvert\vec{z}_1(\mu_1)\rvert$=18.1934 $\lvert\vec{z}_2(\mu_1)\rvert$=13.6393 $\lvert\vec{Z}_3(\mu_1)\rvert$=15.7104	4	2.9817	0.4624

4	$\vec{z}_1(\mu_1)=[10–5.5–16.75]$ $\vec{z}_2(\mu_1)=[7.3041–4.0173–12.2344]$ $\vec{Z}_3(\mu_1)=[8..8923–4.8908–14.8947]$ $\lvert\vec{z}_1(\mu_1)\rvert=20.2685$ $\lvert\vec{z}_2(\mu_1)\rvert=14.8044$ $\lvert\vec{Z}_3(\mu_1)\rvert=18.0234$	4.5	4.5780	0.7433
5	$\vec{z}_1(\mu_1)=[11–6\ 18.5]$ $\vec{z}_2(\mu_1)=[7.8620–4.2884–13.2225]$ $\vec{Z}_3(\mu_1)=[10.0147–5.4625–16.8428]$ $\lvert\vec{z}_1(\mu_1)\rvert=22.3439$ $\lvert\vec{z}_2(\mu_1)\rvert=15.9699$ $\lvert\vec{Z}_3(\mu_1)\rvert=20.3424$	5	4.1743	1.0241
6	$\vec{z}_1(\mu_1)=[12–6.5–20.25]$ $\vec{z}_2(\mu_1)=[8..4206–4.5612–14.2098]$ $\vec{Z}_3(\mu_1)=[11.1375–6.0328–18.7946]$ $\lvert\vec{z}_1(\mu_1)\rvert=24.4195$ $\lvert\vec{z}_2(\mu_1)\rvert=17.1356$ $\lvert\vec{Z}_3(\mu_1)\rvert=22.6644$	5.5	4.7706	1.3050

7	$\vec{z}_1(\mu_1)$=[13–7 22] $\vec{z}_2(\mu_1)$=[8..9786–4.8352–15.1963] $\vec{Z}_3(\mu_1)$=[12.2606–6.6019–20.7487] $\lvert\vec{z}_1(\mu_1)\rvert$=26.4953 $\lvert\vec{z}_2(\mu_1)\rvert$=18.3014 $\lvert\vec{Z}_3(\mu_1)\rvert$=24.9883	6	5.3670	1.5859
8.	$\vec{z}_1(\mu_1)$=[14–7.5–23.75] $\vec{z}_2(\mu_1)$=[9.5391–5.1102–16.1824] $\vec{Z}_3(\mu_1)$=[13.3838–7.1699–22.7047] $\lvert\vec{z}_1(\mu_1)\rvert$=28.5712 $\lvert\vec{z}_2(\mu_1)\rvert$=19.4674 $\lvert\vec{Z}_3(\mu_1)\rvert$=27.3137	6.5	5.9633	1.8668
9	$\vec{z}_1(\mu_1)$=[15–8. 25.5] $\vec{z}_2(\mu_1)$=[10.0989–5.3861–17.1681] $\vec{Z}_3(\mu_1)$=[14.5071–7.7371–24.6621] $\lvert\vec{z}_1(\mu_1)\rvert$=30.6472 $\lvert\vec{z}_2(\mu_1)\rvert$=20.6334 $\lvert\vec{Z}_3(\mu_1)\rvert$=29.6402	7	6.5596	2.1477

10	$\vec{z}_1(\mu_1)$=[16–8..5–27.25] $\vec{z}_2(\mu_1)$=[10.6589–5.6625–18.1534] $\vec{Z}_3(\mu_1)$=[15.6305–8..3037–26.6207] $\lvert\vec{z}_1(\mu_1)\rvert$=32.7233 $\lvert\vec{z}_2(\mu_1)\rvert$=21.7996 $\lvert\vec{Z}_3(\mu_1)\rvert$=31.9675	7.5	7.1560	2.4285
11	$\vec{z}_1(\mu_1)$=[17–9 29] $\vec{z}_2(\mu_1)$=[11.2191–5.9395–19.1385] $\vec{Z}_3(\mu_1)$=[16.7539–8..8697–28.5801] $\lvert\vec{z}_1(\mu_1)\rvert$=34.7995 $\lvert z_2(\mu_1)\rvert$=22.9658 $\lvert\vec{Z}_3(\mu_1)\rvert$=34.2956	8.	7.7523	2.7094

TABLE 3 Numerical values of the ratios of module and appropriate coordinates of predictable points-vectors for different values of parameters $3{,}1769, \le \mu_1 \le 8$ and $0 \le \mu_3 \le 2{,}7094$.

	$\dfrac{\lvert\vec{z}_1(\mu_1)\rvert}{\lvert\vec{Z}_3(\mu_1)\rvert}$	$\dfrac{\lvert\vec{z}_2(\mu_1)\rvert}{\lvert\vec{Z}_3(\mu_1)\rvert}$	$\dfrac{\lvert\vec{z}_1(\mu_1)\rvert}{\lvert\vec{z}_2(\mu_1)\rvert}$
$\mu_1 = 3{,}1769$ $\mu_2 = 2$ $\mu_3 = 0$	1.2647	1	1.2647
	$\vec{z}_2(1)/\vec{Z}_3(1)$	$\vec{z}_2(1)/\vec{Z}_3(1)$	$\vec{z}_1(1)/\vec{z}_2(1)$
	1	1	1.2607

	$\vec{z}_1(2)/\vec{Z}_3(2)$	$\vec{z}_2(2)/\vec{Z}_3(2)$	$\vec{z}_1(2)/\vec{z}_2(2)$
	1.2607	1	1.2607
	$\vec{z}_1(3)/\vec{Z}_3(3)$	$\vec{z}_2(3)/\vec{Z}_3(3)$	$\vec{z}_1(3)/\vec{z}_2(3)$
	1.2659	1	1.2667
	$\lvert\vec{z}_1(\mu_1)\rvert/\lvert\vec{Z}_3(\mu_1)\rvert$	$\lvert\vec{z}_2(\mu_1)\rvert/\lvert\vec{Z}_3(\mu_1)\rvert$	$\lvert\vec{z}_1(\mu_1)\rvert/\lvert\vec{z}_2(\mu_1)\rvert$
	1.2006	0.9292	1.2921
	$\vec{z}_1(1)/\vec{Z}_3(1)$	$\vec{z}_2(1)/\vec{Z}_3(1)$	$\vec{z}_1(1)/\vec{z}_2(1)$
	1.2006	0.9292	1.2921
$\mu_1 = 3,5$ $\mu_2 = 2,3853$ $\mu_3 = 0,1815$	$\vec{z}_1(2)/\vec{Z}_3(2)$	$\vec{z}_2(2)/\vec{Z}_3(2)$	$\vec{z}_1(2)/\vec{z}_2(2)$
	1.2006	0.9282	1.2921
	$\vec{z}_1(3)/\vec{Z}_3(3)$	$\vec{z}_2(3)/\vec{Z}_3(3)$	$\vec{z}_1(3)/\vec{z}_2(3)$
	1.2006	0.9292	1.2921
	$\lvert\vec{z}_1(\mu_1)\rvert/\lvert\vec{Z}_3(\mu_1)\rvert$	$\lvert\vec{z}_2(\mu_1)\rvert/\lvert\vec{Z}_3(\mu_1)\rvert$	$\lvert\vec{z}_1(\mu_1)\rvert/\lvert\vec{z}_2(\mu_1)\rvert$
	1.1580	0.8682	1.3339
	$\vec{z}_1(1)/\vec{Z}_3(1)$	$\vec{z}_2(1)/\vec{Z}_3(1)$	$\vec{z}_1(1)/\vec{z}_2(1)$
$\mu_1 = 4$ $\mu_2 = 2,9817$ $\mu_3 = 0,4624$	1.1580	0.8682	1.3339
	$\vec{z}_1(2)/\vec{Z}_3(2)$	$\vec{z}_2(2)/\vec{Z}_3(2)$	$\vec{z}_1(2)/\vec{z}_2(2)$
	1.1580	0.8682	3.3339
	$\vec{z}_1(3)/\vec{Z}_3(3)$	$\vec{z}_2(3)/\vec{Z}_3(3)$	$\vec{z}_1(3)/\vec{z}_2(3)$
	1.1580	0.8682	1.3339

	$\lvert\vec{z}_1(\mu_1)\rvert/\lvert\vec{Z}_3(\mu_1)\rvert$	$\lvert\vec{z}_2(\mu_1)\rvert/\vec{Z}_3(\mu_1)$	$\lvert\vec{z}_1(\mu_1)\rvert/\lvert\vec{z}_2(\mu_1)\rvert$
$\mu_1 = 4,5$	1.1246	0.8214	1.3691
$\mu_2 = 2,5780$	$\vec{z}_1(1)/\vec{Z}_3(1)$	$\vec{z}_2(1)/\vec{Z}_3(1)$	$\vec{z}_1(1)/\vec{z}_2(1)$
$\mu_3 = 0,7433$	1.1246	0.8214	1.3691
	$\vec{z}_1(2)/\vec{Z}_3(2)$	$\vec{z}_2(2)/\vec{Z}_3(2)$	$\vec{z}_1(2)/\vec{z}_2(2)$
	1.1246	0.8214	1.3691
	$\vec{z}_1(3)/\vec{Z}_3(3)$	$\vec{z}_2(3)/\vec{Z}_3(3)$	$\vec{z}_1(3)/\vec{z}_2(3)$
	1.1246	0.8214	1.3691
	$\lvert\vec{z}_1(\mu_1)\rvert/\lvert\vec{Z}_3(\mu_1)\rvert$	$\vec{z}_2(\mu_1)/\vec{Z}_3(\mu_1)$	$\lvert\vec{z}_1(\mu_1)\rvert/\lvert\vec{z}_2(\mu_1)\rvert$
$\mu_1 = 5$	1.0984	0.7851	1.3991
$\mu_2 = 4,1743$			
$\mu_3 = 1,0241$	$\vec{z}_1(1)/\vec{Z}_3(1)$	$\vec{z}_2(1)/\vec{Z}_3(1)$	$\vec{z}_1(1)/\vec{z}_2(1)$
	1.0984	0.7851	1.3991
	$\vec{z}_1(2)/\vec{Z}_3(2)$	$\vec{z}_2(2)/\vec{Z}_3(2)$	$\vec{z}_1(2)/\vec{z}_2(2)$
	1.0984	0.7851	1.3991
	$\vec{z}_1(3)/\vec{Z}_3(3)$	$\vec{z}_2(3)/\vec{Z}_3(3)$	$\vec{z}_1(3)/\vec{z}_2(3)$
	1.0984	0.7851	1.3991
	$\lvert\vec{z}_1(\mu_1)\rvert/\lvert\vec{Z}_3(\mu_1)\rvert$	$\vec{z}_2(\mu_1)/\vec{Z}_3(\mu_1)$	$\lvert\vec{z}_1(\mu_1)\rvert/\lvert\vec{z}_2(\mu_1)\rvert$
$\mu_1 = 5,5$	1.0774	0.7561	1.4251
$\mu_2 = 4,7706$			
$\mu_3 = 1,30500$			
	$\vec{z}_1(1)/\vec{Z}_3(1)$	$\vec{z}_2(1)/\vec{Z}_3(1)$	$\vec{z}_1(1)/\vec{z}_2(1)$
	1.0774	0.7561	1.4251

	$\vec{z}_1(2)/\vec{Z}_3(2)$	$\vec{z}_2(2)/\vec{Z}_3(2)$	$\vec{z}_1(2)/\vec{z}_2(2)$
	1.0774	0.7561	1.4251
	$\vec{z}_1(3)/\vec{Z}_3(3)$	$\vec{z}_2(3)/\vec{Z}_3(3)$	$\vec{z}_1(3)/\vec{z}_2(3)$
	1.0774	0.7561	1.4251
	$\lvert\vec{z}_1(\mu_1)\rvert/\lvert\vec{Z}_3(\mu_1)\rvert$	$\vec{z}_2(\mu_1)/\vec{Z}_3(\mu_1)$	$\lvert\vec{z}_1(\mu_1)\rvert/\lvert\vec{z}_2(\mu_1)\rvert$
	1.0503	0.7324	1.4477
	$\vec{z}_1(1)/\vec{Z}_3(1)$	$\vec{z}_2(1)/\vec{Z}_3(1)$	$\vec{z}_1(1)/\vec{z}_2(1)$
	1.0603	0.7324	1.4477
$\mu_1=6$			
$\mu_2=5{,}3670$	$\vec{z}_1(2)/\vec{Z}_3(2)$	$\vec{z}_2(2)/\vec{Z}_3(2)$	$\vec{z}_1(2)/\vec{z}_2(2)$
$\mu_3=1{,}5859$	1.0603	0.7324	1.4477
	$\vec{z}_1(3)/\vec{Z}_3(3)$	$\vec{z}_2(3)/\vec{Z}_3(3)$	$\vec{z}_1(3)/\vec{z}_2(3)$
	1.0603	0.7324	1.4477
	$\lvert\vec{z}_1(\mu_1)\rvert/\lvert\vec{Z}_3(\mu_1)\rvert$	$\vec{z}_2(\mu_1)/\vec{Z}_3(\mu_1)$	$\lvert\vec{z}_1(\mu_1)\rvert/\lvert\vec{z}_2(\mu_1)\rvert$
$\mu_1=6{,}5$	1.0460	0.7127	1.4676
$\mu_2=5{,}9633$			
$\mu_3=1{,}8668$	$\vec{z}_1(1)/\vec{Z}_3(1)$	$\vec{z}_2(1)/\vec{Z}_3(1)$	$\vec{z}_1(1)/\vec{z}_2(1)$
	1.0460	0.7127	1.4676
	$\vec{z}_1(2)/\vec{Z}_3(2)$	$\vec{z}_2(2)/\vec{Z}_3(2)$	$\vec{z}_1(2)/\vec{z}_2(2)$
	1.0460	0.7129	1.4676
	$\vec{z}_1(3)/\vec{Z}_3(3)$	$\vec{z}_2(3)/\vec{Z}_3(3)$	$\vec{z}_1(3)/\vec{z}_2(3)$
	1.0460	0.7127	1.4676

	$\lvert\vec{z}_1(\mu_1)\rvert / \lvert\vec{Z}_3(\mu_1)\rvert$ 1.0340	$\vec{z}_2(\mu_1)/\vec{Z}_3(\mu_1)$ 0.6961	$\lvert\vec{z}_1(\mu_1)\rvert / \lvert\vec{z}_2(\mu_1)\rvert$ 1.4853
$\mu_1 = 7$ $\mu_2 = 6{,}5596$ $\mu_3 = 2{,}1477$	$\vec{z}_1(1)/\vec{Z}_3(1)$ 1.0340	$\vec{z}_2(1)/\vec{Z}_3(1)$ 0.6961	$\vec{z}_1(1)/\vec{z}_2(1)$ 1.4853
	$\vec{z}_1(2)/\vec{Z}_3(2)$ 1.0340	$\vec{z}_2(2)/\vec{Z}_3(2)$ 0.6961	$\vec{z}_1(2)/\vec{z}_2(2)$ 1.4853
	$\vec{z}_1(3)/\vec{Z}_3(3)$ 1.0340	$\vec{z}_2(3)/\vec{Z}_3(3)$ 0.6961	$\vec{z}_1(3)/\vec{z}_2(3)$ 1.4853
$\mu_1 = 7{,}5$ $\mu_2 = 7{,}1560$ $\mu_3 = 2{,}4285$	$\lvert\vec{z}_1(\mu_1)\rvert / \lvert\vec{Z}_3(\mu_1)\rvert$ 1.0236	$\vec{z}_2(\mu_1)/\vec{Z}_3(\mu_1)$ 0.6819	$\lvert\vec{z}_1(\mu_1)\rvert / \lvert\vec{z}_2(\mu_1)\rvert$ 1.5011
	$\vec{z}_1(1)/\vec{Z}_3(1)$ 1.0236	$\vec{z}_2(1)/\vec{Z}_3(1)$ 0.6819	$\vec{z}_1(1)/\vec{z}_2(1)$ 1.5011
	$\vec{z}_1(2)/\vec{Z}_3(2)$ 1.0236	$\vec{z}_2(2)/\vec{Z}_3(2)$ 0.6819	$\vec{z}_1(2)/\vec{z}_2(2)$ 1.5011
	$\vec{z}_1(3)/\vec{Z}_3(3)$ 1.0236	$\vec{z}_2(3)/\vec{Z}_3(3)$ 0.6819	$\vec{z}_1(3)/\vec{z}_2(3)$ 1.5011
$\mu_1 = 8$ $\mu_2 = 7{,}7523$ $\mu_3 = 2{,}7094$	$\lvert\vec{z}_1(\mu_1)\rvert / \lvert\vec{Z}_3(\mu_1)\rvert$ 1.0147	$\vec{z}_2(\mu_1)/\vec{Z}_3(\mu_3)$ 0.6696	$\lvert\vec{z}_1(\mu_1)\rvert / \lvert\vec{z}_2(\mu_1)\rvert$ 1.5153
	$\vec{z}_1(1)/\vec{Z}_3(1)$ 1.0147	$\vec{z}_2(1)/\vec{Z}_3(1)$ 0.6696	$\vec{z}_1(1)/\vec{z}_2(1)$ 1.5153

$\vec{z}_1(2)/\vec{Z}_3(2)$	$\vec{z}_2(2)/\vec{Z}_3(2)$	$\vec{z}_1(2)/\vec{z}_2(2)$
1.0147	0.6696	1.5153
$\vec{z}_1(3)/\vec{Z}_3(3)$	$\vec{z}_2(3)/\vec{Z}_3(3)$	$\vec{z}_1(3)/\vec{z}_2(3)$
1.0470	0.6696	1.5153

The data of Tables 2 and 3 allow to conduct deep quality and quantity analysis on prediction of economic event, i.e., to work out numerically all possible variants of prediction data of economic state on the subsequent stage both by total indices on the whole, i.e., by the coordinates of the vectors $|\vec{z}_1(\mu_1)|$, $|\vec{z}_2(\mu_1)|$, $|\vec{Z}_3(\mu_1)|$ and by separate economic factors, i.e., by the coordinates of the vectors $\vec{z}_1(\mu_1)$, $\vec{z}_2(\mu_1)$, $\vec{z}_3(\mu_1)$. Furthermore, it is possible to compare prediction values of economic event by three criteria: (1) by the results of calculations on linear criterion; (2) by the results of calculations according to continuation of the points of the second piecewise-linear vector-function; (3) by the results of calculations of a vector function with regard to uncertainty factors influence. The scheme of comparison of predictable variants are graphically represented in Fig. 1, and in numerical form in Tables 2 and 3. Here for any value of an arbitrary parameter μ_1, changing in the interval $\mu_1^{k_2} \le \mu_1 \le \mu^*$ we have appropriate numerical values $\vec{z}_1(\mu_1)$, $\vec{z}_2(\mu_1)$, $\vec{Z}_3(\mu_1)$, $|\vec{z}_1(\mu_1)|$, $|\vec{z}_2(\mu_1)|$, $|\vec{Z}_3(\mu_1)|$.

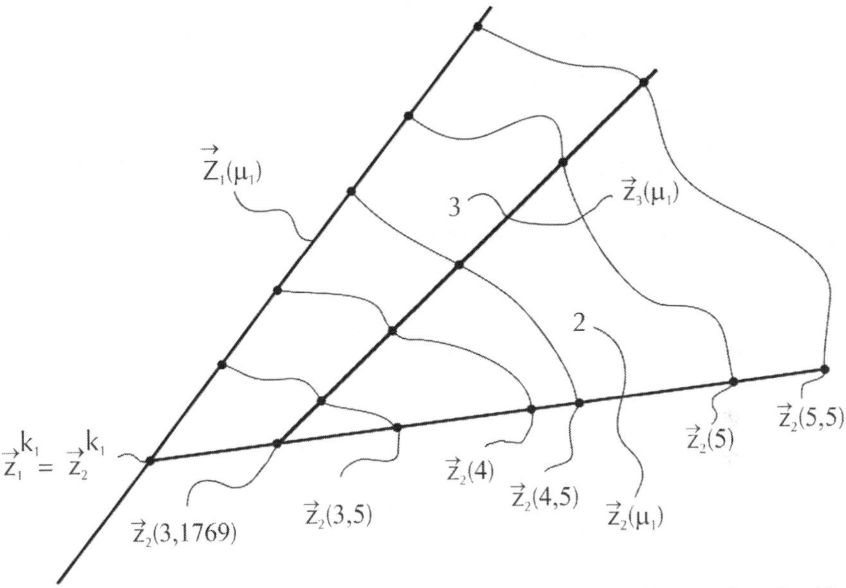

FIGURE 1 The graph of numerical values and appropriate coordinates of predictable points-vectors for the values of the parameters $3{,}1769 \le \mu_1 \le 8$ and $0 \le \mu_3 \le 2{,}7094$ calculated by different criteria.

For visually, as an example we take the value of the parameter $\mu_1 = 5$.

Take into attention the denotation of appropriate ratios of coordinates of vectors $\vec{z}_1(\mu_1)$ $\vec{z}_2(\mu_1)$ $\vec{Z}_3(\mu_1)$ in the form:

$$\frac{\vec{z}_1(i)}{\vec{Z}_3(i)} = \frac{x_{1i}}{X_{3i}}, \quad \frac{\vec{z}_2(i)}{\vec{Z}_3(i)} = \frac{x_{2i}}{X_{3i}}, \quad \frac{\vec{z}_1(i)}{\vec{z}_2(i)} = \frac{x_{1i}}{x_{2i}} \tag{20}$$

In these denotation, compose their percentage ratio (for i=1, 2, 3):

$$n_{1i} = \frac{\vec{z}_1(i)}{\vec{Z}_3(i)} 100\% = \frac{x_{1i}}{X_{3i}} 100\%,$$

$$n_{2i} = \frac{\vec{z}_2(i)}{\vec{Z}_3(i)} 100\% = \frac{x_{2i}}{X_{3i}} 100\%,$$

$$n_{3i} = \frac{\vec{z}_1(i)}{\vec{z}_2(i)} 100\% = \frac{x_{1i}}{x_{2i}} 100\%, \; n_4 = \frac{|\vec{z}_1(\mu_1)|}{|\vec{Z}_3(\mu_1)|} 100\%,$$

$$n_5 = \frac{\left|\vec{z}_2(\mu_1)\right|}{\left|\vec{Z}_3(\mu_1)\right|}100\%, \; n_6 = \frac{\left|\vec{z}_1(\mu_1)\right|}{\left|\vec{z}_2(\mu_1)\right|}100\% \tag{21}$$

According to Eq. (21), Tables 2 and 3, we numerically establish the percentage ratio of the coordinates of the vectors $\vec{z}_1(\mu_1)$ and $\vec{z}_2(\mu_1)$, i.e., x_{1i} and x_{2i} from the appropriate coordinates of the predicting function $\vec{Z}_3(\mu_1)$ with regard to uncertainty factors influence, in the form:

$$n_{11} = n_{12} = n_{13} = 109{,}84\% \tag{22}$$

$$n_{21} = n_{22} = n_{23} = 78{,}51\% \tag{23}$$

$$n_{31} = n_{32} = n_{33} = 139{,}91\% \tag{24}$$

$$n_4 = 1{,}0984, \; n_5 = 0{,}7851, \; n_6 = 1{,}3991 \tag{25}$$

Numerical value Eq. (22) shows that the values of the coordinates of prediction variables calculated by linear criterion is higher by 9.84% than the appropriate prediction coordinates calculated according to the vector function with regard to uncertainty factors influence;

- numerical value Eq. (23) shows that the values of coordinate prediction variables calculated by means of the second piecewise-linear vector function is lower by 21.49% than the appropriate prediction coordinates calculated according to the vector-function with regard to uncertainty factors influence;
- numerical value Eq. (24) shows that the values of coordinates of prediction variables calculated by linear criterion is higher by 39.91% than the appropriate prediction coordinates calculated by means of the second piecewise-linear vector-function;
- numerical value Eq. (25) show the percentage ratio of total indices of vector-functions, i.e., by modules of the vectors $\left|\vec{z}_1(\mu_1)\right|$, $\left|\vec{z}_2(\mu_1)\right|$, $\left|\vec{Z}_3(\mu_1)\right|$, calculated according to different criteria.

It should be noted that by means of numerical data of Table 3 it is easy to establish the dependence of coordinates of prediction vector-function depending on the parameter μ_1, i.e. $\vec{z}_1(i) \sim \mu_1$, $\vec{z}_2(i) \sim \mu_1$ and $\vec{Z}_3(i) \sim \mu_1$.

5.4 DEVELOPMENT OF SOFTWARE FOR COMPUTER MODELING AND MULTIVARIANT PREDICTION OF ECONOMIC EVENT AT UNCERTAINTY CONDITIONS ON THE BASE OF 2-COMPONENT PIECEWISE-LINEAR ECONOMIC-MATHEMATICAL MODELS IN 4-DIMENSIONAL VECTOR SPACE

5.4.1 ACTIONS ALGORITHM FOR COMPUTER MODELING BY CONSTRUCTING 2-COMPONENT PIECEWISE-LINEAR ECONOMIC-MATHEMATICAL MODELS IN 4-DIMENSIONAL VECTOR SPACE

In this section, in 4-dimensional vector space, on the base of the *Matlab* program, we represent a necessary algorithm and numerical calculation method for constructing 2-component piecewise-linear economic-mathematical model, and also consider its realization on a concrete example [10, 11].

According to the theory, [2–4, 9] for the case of 2-component piecewise-linear vector-function in a 4-dimensional vector space, write the main equations and mathematical expressions that are subjected to numerical programming.

Let in 4-dimensional space R_4 be given a statistical table describing the economic process in the form of the points (vectors) $\{a_n\}$. Let these points be represented in the form of adjacent 2-component piecewise-linear vector-equation of the form:

$$\vec{z}_1 = \vec{a}_1 + \mu_1(\vec{a}_2 - \vec{a}_1) \tag{26}$$

$$\vec{z}_2 = \vec{z}_1^{k_1} + \mu_2(\vec{a}_3 - \vec{z}_1^{k_1}) \tag{27}$$

where $\vec{z}_1 = \vec{z}_1(z_{11}, z_{12}, z_{13}, z_{14})$ and $\vec{z}_2 = \vec{z}_2(z_{21}, z_{22}, z_{23}, z_{24})$ are the equations of the first and second piecewise-linear straight lines in 3-dimensional vector space. The vectors $\vec{a}_1(a_{11}, a_{12}, a_{13}, a_{14})$, $\vec{a}_2 = \vec{a}_2(a_{21}, a_{22}, a_{23}, a_{24})$ and $\vec{a}_3 = \vec{a}_3(a_{31}, a_{32}, a_{33}, a_{34})$ are the given points (vectors) in 4-dimensional space, of the form:

$$\vec{a}_1 = a_{11}\vec{i}_1 + a_{12}\vec{i}_2 + a_{13}\vec{i}_3 + a_{14}\vec{i}_4,$$

$$\vec{a}_2 = a_{21}\vec{i}_1 + a_{22}\vec{i}_2 + a_{23}\vec{i}_3 + a_{24}\vec{i}_4$$

$$\vec{a}_3 = a_{31}\vec{i}_1 + a_{32}\vec{i}_2 + a_{33}\vec{i}_3 + a_{34}\vec{i}_4 \tag{28}$$

Here, $\mu_1 \geq 0$ and $\mu_2 \geq 0$ are arbitrary parameters $\vec{z}_1^{k_1}$ is the intersection point of the straight lines $\vec{z}_1$ and $\vec{z}_2$.

The goal of the investigation is the following. Giving the approximate points $\vec{a}_1$, $\vec{a}_2$, $\vec{a}_3$ and also the value of the parameters $\mu_1^{k_1} = \mu_1^*$ and $\mu_2^{k_2} = \mu_2^*$, to work out a computer calculation algorithm of the following equations and mathematical expressions:

$$\vec{z}_1^{k_1} = \vec{a}_1 + \mu_1^{k_1}(\vec{a}_2 - \vec{a}_1)$$

$$\mu_1^{k_2} = \mu_1^{k_1} + \mu_2^{k_2} \frac{(\vec{a}_3 - z_1^{k_1})^2}{(\vec{a}_3 - z_1^{k_1})(\vec{a}_2 - \vec{a}_1)}$$

$$\vec{z}_1^{k_2} = \vec{a}_1 + \mu_1^{k_2}(\vec{a}_2 - \vec{a}_1)$$

$$\vec{z}_2^{k_2} = \vec{z}_1^{k_1} + (\mu_1^{k_2} - \mu_1^{k_1}) \frac{(\vec{a}_3 - \vec{z}_1^{k_1})(\vec{a}_3 - a_1)}{(\vec{a}_3 - \vec{z}_1^{k_1})^2}(\vec{a}_3 - \vec{z}_1^{k_1})$$

$$cos\alpha_{1,2} = \frac{(\vec{z}_1^{k_2} - \vec{z}_1^{k_1})(\vec{z}_1^{k_2} - \vec{z}_1^{k_1})}{\left|\vec{z}_1^{k_2} - \vec{z}_1^{k_1}\right|\left|\vec{z}_1^{k_2} - \vec{z}_1^{k_1}\right|}$$

$$A = (\mu_1^{k_1} - \mu_1^{k_2})\frac{\left|\vec{a}_2 - \vec{a}_1\right|\left|\vec{z}_1^{k_1} - \vec{a}_1\right|}{\vec{z}_1^{k_2}(\vec{z}_1^{k_1} - \vec{a}_1)}$$

$$\lambda_2 = \frac{\mu_2^{k_2}}{\mu_1^{k_1} - \mu_2^{k_2}} \cdot \frac{\left|\vec{z}_1^{k_2} - \vec{z}_1^{k_1}\right|\left|\vec{a}_3 - \vec{z}_1^{k_1}\right|}{\vec{z}_1^{k_2}(\vec{z}_1^{k_2} - \vec{z}_1^{k_1})} \cdot \frac{\vec{z}_1^{k_2}(\vec{z}_1^{k_1} - \vec{a}_1)}{\left|\vec{a}_2 - \vec{a}_1\right|\left|\vec{z}_1^{k_1} - \vec{a}_1\right|}$$

$$\omega_2(\lambda_2, \alpha_{1,2}) = \lambda_2 cos\alpha_{1,2}$$

$$\vec{z}_2 = \vec{z}_1^{k_2}\{1 + A[1 + \omega_2(\lambda_2, \alpha_{1,2})]\}$$

$$\vec{z}_1 = \vec{a}_1 + \mu_1(\vec{a}_2 - \vec{a}_1)$$

$$\mu_2 = (\mu_1 - \mu_1^{k_1}) + \frac{(\vec{a}_3 - \vec{z}_1^{k_1})(\vec{a}_2 - \vec{a}_1)}{(\vec{a}_3 - \vec{z}_1^{k_1})^2}$$

$$\mu_2 = (\mu_1 - \mu_1^{k_1}) + \frac{(\vec{a}_3 - \vec{z}_1^{k_1})(\vec{a}_2 - \vec{a}_1)}{(\vec{a}_3 - \vec{z}_1^{k_1})^2}$$

$$\lambda_2(\mu_1) = \frac{\mu_2}{\mu_1^{k_1} - \mu_1} \cdot \frac{\left|\vec{z}_1 - \vec{z}_1^{k_1}\right|\left|\vec{a}_3 - \vec{z}_1^{k_1}\right|}{\vec{z}_1(\vec{z}_1 - \vec{z}_1^{k_1})} \cdot \frac{\vec{z}_1(\vec{z}_1^{k_1} - \vec{a}_1)}{\left|\vec{a}_2 - \vec{a}_1\right|\left|\vec{z}_1^{k_1} - \vec{a}_1\right|}$$

$$\omega_2(\mu_1) = \lambda_2(\mu_1)cos\alpha_{1,2}$$

$$\vec{z}_2(\mu_1) = \vec{z}_1\{1 + A\ (\mu_1)[1 + \omega_2(\mu_1)]\} \tag{29}$$

Introduce the following denotation:

$$\begin{aligned}
&|\vec{a}_1 \to a1;\ \vec{a}_2 \to a2;\ \vec{a}_3 \to a3;\ \mu_1 \to m1;\ \mu_1^{k_1} \to m1k1 \\
&\mu_1^{k_2} \to m1k2;\ \vec{z}_1^{k_1} \to z1k1,\ \vec{z}_1^{k_2} \to z1k2;\ \vec{z}_2^{k_2} \to z2k2 \\
&\vec{z}_1 \to z1;\ \mu_2 \to m2,\ \mu_2^{k_2} \to m2k2,\ A(\mu_1) \to Am1; \\
&\lambda_2 \to La2,\ \lambda_2(\mu_1) \to La2m1; \\
&\omega(\lambda_2, \alpha_{12}) = \omega_2(\mu_1) \to w2m1;\ \vec{z}_2 \to z2, \\
&\vec{z}_2(\mu_1) \to z2m1
\end{aligned} \tag{30}$$

Using the introduced denotation Eq. (30), we compose a program for numerical construction of 2-component piecewise-linear economic-mathematical models with regard to unaccounted factors influence in 4-dimensional vector space in the *Matlab* program, in the following form:

a_1=[a_{11} a_{12} a_{13} a_{14}]

a_2=[a_{21} a_{22} a_{23} a_{24}]

a_3=[a_{31} a_{32} a_{33} a_{34}]

m_1k_1=(m_1)*m_2k_2=(m_2)*

for m_1=J_1: j_2:J_3

z_1k_1=a_1+m_1k_1*(a_2–a_1);

m_1k_2=m_1k_1+m_2k_2*((a_3–z_1k_1)*

(a_3–z_1k_1)’)/((a_3–z_1k_1)*(a_2–a_1)’);

z_1k_2=a_1+m_1k_2*(a_2–a_1);

z_2k_2=z_1k_1+(m_1k_2–m_1k_1)*((a_3–z_1k_1)*

$(a_2-a_1)')/((a_3-z_1k_1)*(a_3-z_1k_1)')*(a_3-z_1k_1);$

$cosa_{12}=((z_1k_2-z_1k_1)*(z_2k_2-z_1k_1)')/(sqrt((z_1k_2-z_1k_1)$

$*(z_1k_2-z_1k_1)')*sqrt((z_2k_2-z_1k_1)*(z_2k_2-z_1k_1)'))$

$A=(m_1k_1-m_1k_2)*(sqrt((a_2-a_1)*(a_2-a_1)')*sqrt((z_1k_1-a_1)*$

$(z_1k_1-a_1)'))/(z_1k_2*(z_1k_1-a_1)');$

$p_1=m_2k_2/(m_1k_1-m_1k_2);$

$p_2=(sqrt((z_1k_2-z_1k_1)*(z_1k_2-z_1k_1)')*sqrt((a_3-z_1k_1)*$

$(a_3-z_1k_1)'))/(z_3)');$

$p_3=(z_1k_2*(z_1k_1-a_1)')/(sqrt((a_2-a_1)*(a_2-a_1)')*$

$sqrt((z_1k_1-a_1)*(z_1k_1-a_1)'));$

$La_2=p_1*p_2*p_3;\ w_2=La_2*cosa_{12};$

$z_2=z_1k_2*(1+A*(1+w_2));$

$z_1=a_1+m_1*(a_2-a_1)$

$m_2=(m_1-m_1k_1)*(((a_3-z_1k_1)*(a_2-a_1)')/((a_3-z_1k_1)*$

$(a_3-z_1k_1)'))$

$Am_1=(m_1k_1-m_1)*(sqrt((a_2-a_1)*(a_2-a_1)')*$

$sqrt((z_1k_1-a_1)*(z_1k_1-a_1)'))/(z_1*(z_1k_1-a_1)')$

$p_1\ m_1=m_2/(m_1k_1-m_1);$

$p_2\ m_1=(sqrt((z_1-z_1k_1)*(z_1-z_1k_1)')*sqrt((a_3-z_1k_1)*$

$(a_3-z_1k_1)'))/(z_1*(z_1-z_1k_1)');$

$p_3\ m_1=(z_1*(z_1k_1-a_1)')/(sqrt((a_2-a_1)*(a_2-a_1)')*$

$sqrt((z_1k_1-a_1)*(z_1k_1-a_1)'));$

$La_2\ m_1=p_1\ m_1*p_2\ m_1*p_3\ m_1;$

$w_2\ m_1=La_2\ m_1*cosa_{12}$

$z_2\ m_1=z_1*(1+Am_1*(1+w_2\ m_1))$

end (5.4.6)

Example. Consider the following table of statistical data. Let in 4-dimensional space be given the vectors $\vec{a}_1$, $\vec{a}_2$, $\vec{a}_3$, parameters $\mu_1^{k_1}$ and $\mu_2^{k_2}$ in the following numerical form:

a_1=[1–1 1–1]; a_2=[3–2 4, 5–5];
a_3=[6–4 7–6];
m_1k_1=1.5
m_2k_2=2
for m_1=1,5:0,5:8.

The task of the investigation is to represent the points of the second piecewise-linear straight line depending on the first piecewise-linear vector function $\vec{z}_1(\mu_1)$ and the unaccounted factors influence function $\omega_2(\lambda_2, \alpha_{1,2})$ for arbitrary values of the parameter μ_1 changing in the interval $\mu_1^{k_1} = 1,5 \le \mu_1 \le \mu_1^* = 8$, in the form:

$$\vec{z}_2(\mu_1) = \vec{z}_1(\mu_1)\{1 + A(\mu_1)[1 + \omega_2(\mu_1)]\}$$

Below, in Table 4 we represent the appropriate equations of the second piecewise-linear straight line $z_2(\mu_1)$ in 4-dimensional vector space.

It is should be noted that the numerically constructed vectors $\vec{z}_2(\mu_1)$ for different values of the parameter $1,5 \le \mu_1 \ge 8$ in 4-dimensional space completely coincides with numerical results obtained earlier in Refs. [2–4, 9] that were stated in detail in Chapter 3.

5.4.2 ALGORITHM OF MULTIVARIANT COMPUTER MODELING OF PREDICTION VARIABLES OF ECONOMIC EVENT IN 4-DIMENSIONAL VECTOR SPACE

Based around the *Matlab* program, we suggested an algorithm and numerical program for investigating multivariant prediction of economic event at numerical conditions on the base of 2-component piecewise-linear economic-mathematical model in 4-dimensional vector space [2–4, 9].

According to the theory [7, 8] for the case of 2-component piecewise-linear vector function at uncertainty conditions in 4-dimensional vector

Table 5.4 Numerical values of the points (vectors) of the second piecewise-linear straight line $\Xi_2(\mu_1)$ in 4-dimensional vector space vector space depending on the parameter $\mu_1 \geq 1{,}5$, defined by the formula: $\vec{\Xi}_2(\mu_1) = \vec{\Xi}_1(\mu_1)\{1 + A(\mu_1)[1 + \omega_2(\mu_1)]\}$, for the following given values of the vectors $\vec{a}_1 = \vec{i}_1 + \vec{i}_2 + \vec{i}_3 + \vec{i}_4$, $\vec{a}_2 = 3\vec{i}_1 + 2\vec{i}_2 + 4{,}5\vec{i}_3 + 5\vec{i}_4$, $\vec{a}_3 = 6\vec{i}_1 + 4\vec{i}_2 + 7\vec{i}_3 + 6\vec{i}_4$, and also the parameters $\mu_1^{k_1} = 1{,}5$ and $\mu_2^{k_2} = 2$

N	μ_1	μ_2	$A(\mu_1)$	$\omega_2(\mu_1)$	$\vec{\Xi}_1(\mu_1) = \vec{a}_1 + \mu_1(\vec{a}_2 - \vec{a}_1)$	$\vec{\Xi}_2(\mu_1) = \vec{\Xi}_1(\mu_1)\{1 + A(\mu_1)[1 + \omega_2(\mu_1)]\}$
1	1,5	0	0	-0,0655	$\vec{\Xi}_1$(1,5)=[4 2,5 6,25 7]	$\vec{\Xi}_2$(1,5)=[4 2,5 6,25 7]
2	2	0,2640	-0,2159	-0,0655	$\vec{\Xi}_1$(2)=[5 3 8 9]	$\vec{\Xi}_2$(2)=[4,539 2,7234 7,2625 7,1841]
3	2,5	0,5280	-0,3551	-0,0655	$\vec{\Xi}_1$(2,5)=[6 3,5 9,75 11]	$\vec{\Xi}_2$(2,5)=[5,086 2,9668 8,2647 7,3494]
4	3	0,7920	-0,4524	-0,0655	$\vec{\Xi}_1$(3)=[7 4 11,5 13]	$\vec{\Xi}_2$(3)=[5,6372 3,2213 9,2613 7,5043]
5	3,5	1,0560	-0,5241	-0,0655	$\vec{\Xi}_1$(3,5)=[8 4,5 13,25 15]	$\vec{\Xi}_2$(3,5)=[6,1913 3,4826 10,2544 7,6529]
6	4	1,3200	-0,5793	-0,0655	$\vec{\Xi}_1$(4)=[9 5 15 17]	$\vec{\Xi}_2$(4)=[6,7471 3,7484 11,2452 7,7975]
7	4,5	1,5840	-0,6230	-0,0655	$\vec{\Xi}_1$(4,5)=[10 5,5 16,75 19]	$\vec{\Xi}_2$(4,5)=[7,3041 4,0173 12,2344 7,9392]
8	5	1,8480	-0,6584	-0,0655	$\vec{\Xi}_1$(5)=[11 6 18,5 21]	$\vec{\Xi}_2$(5)=[7,862 4,2884 13,2225 8,0790]
9	5,5	2,1120	-0,6878	-0,0655	$\vec{\Xi}_1$(5,5)=[12 6,5 20,25 23]	$\vec{\Xi}_2$(5,5)=[8,4206 4,5612 14,2098 8,2172]
10	6	2,3760	-0,7125	-0,0655	$\vec{\Xi}_1$(6)=[13 7 22 25]	$\vec{\Xi}_2$(6)=[8,9796 4,8352 15,1963 8,3543]
11	6,5	2,6400	-0,7336	-0,0655	$\vec{\Xi}_1$(6,5)=[14 7,5 23,75 27]	$\vec{\Xi}_2$(6,5)=[9,5391 5,1102 16,1824 8,4905]
12	7	2,9040	-0,7518	-0,0655	$\vec{\Xi}_1$(7)=[15 8 25,5 29]	$\vec{\Xi}_2$(7)=[10,0989 5,3861 17,1681 8,6260]
13	7,5	3,1680	-0,7677	-0,0655	$\vec{\Xi}_1$(7,5)=[16 8,5 27,25 31]	$\vec{\Xi}_2$(7,5)=[10,6589 5,6625 18,1534 8,7609]
14	8	3,4320	-0,7816	-0,0655	$\vec{\Xi}_1$(8)=[17 9 29 33]	$\vec{\Xi}_2$(8)=[4,5824 2,4226 7,8171 8,8953]

space, we have the following equations and expressions (see the equations of Chapters 2 and 3):

$$\vec{z}_1^{k_1} = \vec{a}_1 + \mu_1^{k_1}(\vec{a}_2 - \vec{a}_1)$$

$$\mu_1^{k_2} = \mu_1^{k_1} + \mu_2^{k_2}\frac{(\vec{a}_3 - \vec{z}_1^{k_1})^2}{(\vec{a}_3 - \vec{z}_1^{k_1})(\vec{a}_2 - \vec{a}_1)}$$

$$\vec{z}_1^{k_2} = \vec{a}_1 + \mu_1^{k_2}(\vec{a}_2 - \vec{a}_1)$$

$$\vec{z}_2^{k_2} = \vec{z}_1^{k_1} + (\mu_1^{k_2} - \mu_1^{k_1})\frac{(\vec{a}_3 - \vec{z}_1^{k_1})(\vec{a}_2 - \vec{a}_1)}{(\vec{a}_3 - \vec{z}_1^{k_1})^2}(\vec{a}_3 - \vec{z}_1^{k_1})$$

$$cos\alpha_{1,2} = \frac{(\vec{z}_1^{k_2} - \vec{z}_1^{k_1})(\vec{z}_2^{k_2} - \vec{z}_1^{k_1})}{\left|\vec{z}_1^{k_2} - \vec{z}_1^{k_1}\right| \cdot \left|\vec{z}_2^{k_2} - \vec{z}_1^{k_1}\right|}$$

$$A(\mu_1^{k_2}) = A = (\mu_1^{k_1} - \mu_1^{k_2})\frac{\left|\vec{a}_2 - \vec{a}_1\right|\left|\vec{z}_1^{k_1} - \vec{a}_1\right|}{\vec{z}_1^{k_2}(\vec{z}_1^{k_1} - \vec{a}_1)}$$

$$\lambda_2(\mu_1^{k_2}) = \lambda_2^{k_2} = \frac{\mu_2^{k_2}}{\mu_1^{k_1} - \mu_1^{k_2}} \cdot \frac{\left|\vec{z}_1^{k_2} - \vec{z}_1^{k_1}\right|\left|\vec{a}_3 - \vec{z}_1^{k_1}\right|}{\vec{z}_1^{k_2}(\vec{z}_1^{k_2} - \vec{z}_1^{k_1})} \cdot \frac{\vec{z}_1^{k_2}(\vec{z}_1^{k_1} - \vec{a}_1)}{\left|\vec{a}_2 - \vec{a}_1\right|\left|\vec{z}_1^{k_1} - \vec{a}_1\right|}$$

$$\omega_2(\lambda_2^{k_2}, \alpha_{1,2}) = \lambda_2^{k_2} cos\alpha_{1,2}$$

$$\vec{z}_2(\mu_1^{k_2}) = \vec{z}_2 = \vec{z}_1^{k_2}\{1 + A\,[1 + \omega_2(\lambda_2^{k_2}, \alpha_{1,2})]\}$$

$$\vec{z}_1(\mu_1) = \vec{z}_1 = \vec{a}_1 + \mu_1(\vec{a}_2 - \vec{a}_1)$$

$$\mu_2 = (\mu_1 - \mu_1^{k_1}) \cdot \frac{(\vec{a}_3 - \vec{z}_1^{k_1})(\vec{a}_2 - \vec{a}_1)}{(\vec{a}_3 - \vec{z}_1^{k_1})^2}$$

$$A(\mu_1) = (\mu_1^{k_1} - \mu_1) \frac{|\vec{a}_2 - \vec{a}_1| |\vec{z}_1^{k_1} - \vec{a}_1|}{\vec{z}_1(\vec{z}_1^{k_1} - \vec{a}_1)}$$

$$\lambda_2(\mu_1) = \frac{\mu_2}{\mu_1^{k_1} - \mu_1} \cdot \frac{|\vec{z}_1 - \vec{z}_1^{k_1}| |\vec{a}_3 - \vec{z}_1^{k_1}|}{\vec{z}_1(\vec{z}_1 - \vec{z}_1^{k_1})} \cdot \frac{\vec{z}_1(\vec{z}_1^{k_1} - \vec{a}_1)}{|\vec{a}_2 - \vec{a}_1| |\vec{z}_1^{k_1} - \vec{a}_1|}$$

$$\omega_2(\lambda_2(\mu_1), \alpha_{12}) = \omega_2(\mu_1) = \lambda_2(\mu_1) cos\alpha_{1,2}$$

$$\vec{z}_2(\mu_1) = \vec{z}_1\{1 + A\ (\mu_1)[1 + \omega_2(\mu_1)]\}$$

$$a_{42}(1) = (\vec{a}_4)_2 = z_{22}^{k_2} + \frac{a_{22} - a_{12}}{a_{21} - a_{11}}(a_{41}(1) - z_{21}^{k_2})$$

$$a_{43}(1) = (\vec{a}_4)_3 = z_{23}^{k_2} + \frac{a_{23} - a_{13}}{a_{21} - a_{11}}(a_{41}(1) - z_{21}^{k_2})$$

$$a_{44}(1) = (\vec{a}_4)_4 = z_{24}^{k_2} + \frac{a_{24} - a_{14}}{a_{21} - a_{11}}(a_{41}(1) - z_{21}^{k_2})$$

$$q_1 = \frac{(\vec{a}_2 - \vec{a}_1)(\vec{a}_3 - \vec{z}_1^{k_1})}{(\vec{a}_3 - \vec{z}_1^{k_1})^2} \quad q_2 = \frac{(\vec{a}_3 - \vec{z}_1^{k_1})(\vec{a}_4(1) - \vec{z}_2^{k_2})}{(\vec{a}_4(1) - \vec{z}_2^{k_2})^2}$$

$$\mu_3 = (\mu_1 - \mu_1^{k_2}) q_1 q_2$$

$$q_3 = \frac{\left|\vec{z}_2(\mu_1) - \vec{z}_2^{k_2}\right| \cdot \left|\vec{a}_4(1) - \vec{z}_2^{k_2}\right|}{\vec{z}_1(\vec{z}_2(\mu_1) - \vec{z}_2^{k_2})}$$

$$q_4 = \frac{\vec{z}_1(\vec{z}_1^{k_1} - \vec{a}_1)}{\left|\vec{a}_2 - \vec{a}_1\right| \cdot \left|\vec{z}_1^{k_1} - \vec{a}_1\right|} \quad \lambda_3(\mu_1) = \frac{\mu_3}{\mu_1^{k_1} - \mu_1} \cdot q_3 \cdot q_4$$

$$cos\alpha_{2,3} = \frac{(\vec{z}_2(\mu_1) - \vec{z}_2^{k_2}) \cdot (\vec{a}_4(1) - \vec{z}_2^{k_2})}{\left|\vec{z}_2(\mu_1) - \vec{z}_2^{k_2}\right| \left|\vec{a}_4(1) - \vec{z}_2^{k_2}\right|}$$

$$A_3(\mu_1) = (\mu_1^{k_1} - \mu_1) \cdot \frac{\left|\vec{a}_2 - \vec{a}_1\right| \cdot \left|\vec{z}_1^{k_1} - \vec{a}_1\right|}{\vec{z}_1(\vec{z}_1^{k_1} - \vec{a}_1)}$$

$$\Omega_3(\lambda_3(\mu_1), \alpha_{2,3}) = \lambda_3(\mu_1) \cdot cos\alpha_{2,3}$$

$$\vec{Z}_3(\mu_1) = \vec{z}_1\{1 + A_3(\mu_1)[1 + \omega_2(\mu_1) + \Omega_3(\lambda_3, \alpha_{2,3})]\} \tag{32}$$

Give the approximate points $\vec{a}_1$, $\vec{a}_2$, $\vec{a}_3$, $\vec{a}_4(1)$ and also the values of the parameters $\mu_1^{k_1} = \mu_1^*$ and $\mu_2^{k_2} = \mu_2^*$. Introduce the denotation:

$\vec{a}_1 \to a1$; $\vec{a}_2 \to a2$; $\vec{a}_3 \to a3$; $(\vec{a}_4)_1 \to a4(1)$; $\mu_1 \to m1$;

$\mu_1^{k_1} \to m1k1$; $\mu_1^{k_2} \to m1k2$; $\mu_2 \to m2$; $\mu_2^{k_2} \to m2k2$;

$z_1^{k_1} \to z1k1$; $z_1^{k_2} \to z1k2$; $z_2^{k_2} \to z2k2$; $cos\alpha_{12} \to cosa12$;

$A(\mu_1^{k_2}) = A \to A$; $\lambda_2(\mu_1^{k_2}) = \lambda_2^{k_2} \to La2$;

$\omega_2(\lambda_2^{k_2}, \alpha_{12}) \to w2$; $\vec{z}_2(\mu_1^{k_2}) \to z2$; $\vec{z}_1(\mu_1) \to z1$;

$|\vec{z}_1(\mu_1)| \to z1M$; $A(\mu_1) \to Am1$; $\lambda_2(\ \mu_1) \to La2m1$;

$\omega_2(\lambda_2(\mu_1), \alpha_{12}) \to w2m1$; $\vec{z}_2(\mu_1) \to z2m1$;

$|\vec{z}_2(\mu_1)| \to z2m1M$; $(\vec{a}_4)_2 \to a4(2)$; $(\vec{a}_4)_3 \to a4(3)$;

$(\vec{a}_4)_4 \to a4(4)$; $q_1 = q1$; $q_2 = q2$ $\mu_3 \to m3$;

$\lambda_3(\ \mu_1) \to La3m$; $q_3 = q3$; $q_4 = q4$;

$cos\alpha_{23} \to cosa23$; $A_3(\mu_1) \to Am1p$;

$\Omega_3(\lambda_3(\mu_1), \alpha_{23}) \to w3mp$; $\vec{Z}_3(\mu_1) \to z3m1p$;

$|\vec{Z}_3(\mu_1)| \to z3m1pM$;

$|\vec{z}_1(\mu_1)/\vec{Z}_3(\mu_1)| \to (z1M)/(z3m1pM) = B_1$;

$|\vec{z}_2(\mu_1)/\vec{Z}_3(\mu_1)| \to (z2m1M)/(z3m1pM) = B_2$;

$|\vec{z}_1(\mu_1)/\vec{z}_2(\mu_1)| \to (z1M)/(z2m1M) = B_3$;

$\vec{z}_1(1)/\vec{Z}_3(1) \to (z1(1))/(z3m1p(1)) = B_4$;

$$\begin{aligned}
&\vec{z}_2(1)/\vec{Z}_3(1) \rightarrow (z2m1(1))/(z3m1p(1)) = B_5;\\
&\vec{z}_1(1)/\vec{z}_2(1) \rightarrow (z1(1))/(z2m1(1)) = B_6;\\
&\vec{z}_1(2)/\vec{Z}_3(2) \rightarrow (z1(2))/(z3m1p(2)) = B_7;\\
&\vec{z}_2(2)/\vec{Z}_3(2) \rightarrow (z2m1(2))/(z3m1p(2)) = B_8;\\
&\vec{z}_1(2)/\vec{z}_2(2) \rightarrow (z1(2))/(z2m1(2)) = B_9;\\
&\vec{z}_1(3)/\vec{Z}_3(3) \rightarrow (z1(3))/(z3m1p(3)) = B_{10};\\
&\vec{z}_2(3)/\vec{Z}_3(3) \rightarrow (z2m1(3))/(z3m1p(3)) = B_{11};\\
&\vec{z}_1(3)/\vec{z}_2(3) \rightarrow (z1(3))/(z2m1(3)) = B_{12};\\
&\vec{z}_1(4)/\vec{Z}_3(4) \rightarrow (z1(4))/(z3m1p(4)) = B_{13};\\
&\vec{z}_2(4)/\vec{Z}_3(4) \rightarrow (z2m1(4))/(z3m1p(4)) = B_{14};\\
&\vec{z}_1(4)/\vec{z}_2(4) \rightarrow (z1(4))/(z2m1(4)) = B_{15}
\end{aligned} \tag{33}$$

Using the introduced denotation Eq. (33), an algorithm and appropriate numerical program for the system Eq. (32) in the *Matlab* language will be represented in the form [8, 9]:

a_1=[a_{11} a_{12} a_{13} a_{14}]

a_2=[a_{21} a_{22} a_{23} a_{24}]

a_3=[a_{31} a_{32} a_{33} a_{34}]

m_1k_1=(m_1)*m_2k_2=(m_2)*

a_4(1)=a_4(1)* for m_1=J_1:J_2:J_3

z_1k_1=a_1+m_1k_1*(a_2–a_1);

m_1k_2=m_1k_1+m_2k_2*((a_3–z_1k_1)*

$(a_3-z_1k_1)')/((a_3-z_1k_1)*(a_2-a_1)');$

$z_1k_2=a_1+m_1k_2*(a_2-a_1);$

$z_2k_2=z_1k_1+(m_1k_2-m_1k_1)*((a_3-z_1k_1)*$

$(a_2-a_1)')/((a_3-z_1k_1)*(a_3-z_1k_1)')*(a_3-z_1k_1);$

$cosa_{12}=((z_1k_2-z_1k_1)*(z_2k_2-z_1k_1)')/(sqrt((z_1k_2-z_1k_1)*$

$(z_1k_2-z_1k_1)')*sqrt((z_2k_2-z_1k_1)*(z_2k_2-z_1k_1)'));$

$A=(m_1k_1-m_1k_2)*(sqrt((a_2-a_1)*(a_2-a_1)')*sqrt((z_1k_1-a_1)*$

$(z_1k_1-a_1)'))/(z_1k_2*(z_1k_1-a_1)');$

$p_1=m_2k_2/(m_1k_1-m_1k_2);$

$p_2=(sqrt((z_1k_2-z_1k_1)*(z_1k_2-z_1k_1)')*sqrt((a_3-z_1k_1)*$

$(a_3-z_1k_1)'))/(z_1k_2*(z_1k_2-z_1k_1)');$

$p_3=(z_1k_2*(z_1k_1-a_1)')/(sqrt((a_2-a_1)*(a_2-a_1)')*$

$sqrt((z_1k_1-a_1)*(z_1k_1-a_1)'));$

$La_2=p_1*p_2*p_3;\ w_2=La_2*cosa_{12};$

$z_2=z_1k_2*(1+A*(1+w_2))$

$z_1=a_1+m_1*(a_2-a_1)$

$z_1\ M=sqrt((z_1)*(z_1)')$

$m_2=(m_1-m_1k_1)*(((a_3-z_1k_1)*(a_2-a_1)')/((a_3-z_1k_1)*$

$(a_3-z_1k_1)'))$

$Am_1=(m_1k_1-m_1)*(sqrt((a_2-a_1)*(a_2-a_1)')*sqrt((z_1k_{1-1})*$

$(z_1k_1-a_1)'))/(z_1*(z_1k_1-a_1)');\ p_1\ m_1=m_2/(m_1k_1-m_1);$

$p_2\, m_1=(sqrt((z_1-z_1k_1)*(z_1-z_1k_1)')*sqrt((a_3-z_1k_1)*$

$(a_3-z_1k_1)'))/(z_1*(z_1-z_1k_1)');$

$p_3\, m_1=(z_1*(z_1k_1-a_1)')/(sqrt((a_2-a_1)*(a_2-a_1)')*$

$sqrt((z_1k_1-a_1)*(z_1k_1-a_1)'));$

$La_2\, m_1=p_1\, m_1*p_2\, m_1*p_3\, m_1;$

$w_2\, m_1=La_2\, m_1*cosa_{12};$

$z_2\, m_1=z_1*(1+Am_1*(1+w_2\, m_1))$

$z_2\, m_1\, M=sqrt((z_2\, m_1)*(z_2\, m_1)')$

$a_4(2)=z_2k_2(2)+[(a_2(2)-a_1(2))/(a_2(1)-a_1(1))]*$

$(a_4(1)-z_2k_2(1));$

$a_4(3)=z_2k_2(3)+[(a_2(3)-a_1(3))/(a_2(1)-a_1(1))]*$

$(a_4(1)-z_2k_2(1));$

$a_4(4)=z_2k_2(4)+[(a_2(4)-a_1(4))/(a_2(1)-a_1(1))]*(a_4(1)-z_2k_2(1));$

$q_1=[(a_2-a_1)*(a_3-z_1k_1)']/[(a_3-z_1k_1)*(a_3-z_1k_1)'];$

$q_2=((a_3-z_1k_1)*(a_4-z_2k_2)')/((a_4-z_2k_2)*(a_4-z_2k_2)');$

$m_3=(m_1-m_1k_2)*q_1*q_2$

$q_3=(sqrt((z_2\, m_1-z_2k_2)*(z_2\, m_1-z_2k_2)')*sqrt((a_4-z_2k_2)*$

$(a_4-z_2k_2)'))/(z_1*(z_2\, m_1-z_2k_2)');$

$q_4=[z_1*(z_1k_1-a_1)']/[sqrt((a_2-a_1)*(a_2-a_1)')*$

$sqrt((z_1k_1-a_1)*(z_1k_1-a_1)')];$

$La_3\, m=[m_3/(m_1k_1-m_1)]*q_3*q_4;$

$cosa_{23}=((z_2\ m_1-z_2k_2)*(a_4-z_2k_2)')/[sqrt((z_2\ m_1-z_2k_2)*$

$(z_2\ m_1-z_2k_2)')*sqrt((a_4-z_2k_2)*(a_4-z_2k_2)')];$

$Am_1p=(m_1k_1-m_1)*(sqrt((a_2-a_1)*(a_2-a_1)')*sqrt((z_1k_1-a_1)*$

$(z_1k_1-a_1)'))/(z_1*(z_1k_1-a_1)');$

$w_3\ mp=La_3\ m*cosa_{23};$

$z_3\ m_1p=z_1*[1+Am_1p*(1+w_2\ m_1+w_3\ mp)]$

$z_3\ m_1pM=sqrt((z_3\ m_1p)*(z_3\ m_1p)')$

$B_1=(z_1\ M)/(z_3\ m_1pM)$

$B_2=(z_2\ m_1\ M/(z_3\ m_1pM))$

$B_3=(z_1\ M)/(z_2\ m_1\ M)$

$B_4=(z_1(1))/(z_3\ m_1p(1))$

$B_5=(z_2\ m_1(1))/(z_3\ m_1p(1))$

$B_6=(z_1(1))/(z_2\ m_1(1))$

$B_7=(z_1(2))/(z_3\ m_1p(2))$

$B_8=(z_2\ m_1(2))/(z_3\ m_1p(2))$

$B_9=(z_1(2))/(z_2\ m_1(2))$

$B_{10}=(z_1(3))/(z_3\ m_1p(3))$

$B_{11}=(z_2\ m_1(3))/(z_3\ m_1p(3))$

$B_{12}=(z_1(3))/(z_2\ m_1(3))$

$B_{13}=(z_1(4))/(z_3\ m_1p(4))$
$B_{14}=(z_2\ m_1(4))/(z_3\ m_1p(4))$
$B_{15}=(z_1(4))/(z_2\ m_1(4))$
end (5.4.9)

Giving the statistical data of the vectors $\vec{a}_1$, $\vec{a}_2$, $\vec{a}_3$, $\vec{a}_4(1)$ and parameters $\mu_1^{k_1}$ and $\mu_2^{k_2}$, by means of the above-suggested numerical program we can investigate multivariant prediction economic event at uncertainty conditions on the base of 2-component piecewise-linear model in 4-dimensional vector space.

Example. Consider the case with the following given statistical vectors $\vec{a}_1$, $\vec{a}_2$, $\vec{a}_3$, $\vec{a}_4(1)$ and parameters $\mu_1^{k_1}$ and $\mu_2^{k_2}$:

a_1=[1–1 1–1]

a_2=[3–2 4.5–5]

a_3=[6–4 7–6]

m_1k_1=1.5

m_2k_2=2

$a_4(1)$=10

for m_1=1,5:0,5:8.

For these data, carry out appropriate numerical calculation an establishing all-possible variants of prediction data of economic state on the subsequent stage.

The data of Tables 5 and 6 allow to give quality and quantity analysis on prediction of economic event, i.e., to work out numerically the variants of prediction data of economic state on the subsequent stage both on data indices on the whole, i.e., by the modulus of the vectors $|\vec{z}_1(\mu_1)|$, $|\vec{z}_2(\mu_1)|$, $|\vec{z}_3(\mu_1)|$ and by separate economic factors, i.e., by the coordinates $\vec{z}_1(\mu_1)$, $\vec{z}_2(\mu_1)$, $\vec{z}_3(\mu_1)$ vectors. Furthermore, we can compare the prediction values of economic event by three criteria: 1) by the results of calculations on linear criterion; 2) by the results of calculations according to continuation of the points of the second piecewise-linear vector function; 3) by the results of calculations of vector-function with regard to uncertainty factors influence. The scheme of such a comparison of predictable data are graphically represented in Fig. 2, in numerical form in Tables 5 and 6.

TABLE 5 Numerical values of module and appropriate coordinates of predictable points-vectors for different values of parameters $5{,}2879 \le \mu_1 \le 8$ and $0 \le \mu_3 \le 0{,}1777$ in 4-dimensional vector space.

N	Numerical values of the vectors $\vec{z}_1 \cdot \vec{z}_2 \cdot \vec{z}_3$ and their module	μ_1	μ_2	μ_3
1	$\vec{z}_1(\mu_1)$=[11.5758–6.2879–19.5076–22.1516] $\vec{z}_2(\mu_1)$=[4.2635–2.3159–7.1849–8..1587] $\vec{Z}_3(\mu_1)$=[4.2635–2.3159–7.1849–8..1587] $\lvert\vec{z}_1(\mu_1)\rvert$=32.3230 $\lvert\vec{z}_2(\mu_1)\rvert$=11.9050 $\lvert\vec{Z}_3(\mu_1)\rvert$=11.9050	5.2879	2	0
2	$\vec{z}_1(\mu_1)$=[12–6.5–20.25–23] $\vec{z}_2(\mu_1)$=[4.2872–2.3223–7.2347–8..2172] $\vec{Z}_3(\mu_1)$=[4.2405–2.2969–7.1558–8..1276] $\lvert\vec{z}_1(\mu_1)\rvert$=33.5457 $\lvert\vec{z}_2(\mu_1)\rvert$=11.9848 $\lvert\vec{Z}_3(\mu_1)\rvert$=11.8541	5.5	2.1120	0.0139
3	$\vec{z}_1(\mu_1)$=[13–7 22–25] $\vec{z}_2(\mu_1)$=[4.3442–2.3392–7.3518–8..3543] $\vec{Z}_3(\mu_1)$=[4.5169–2.4322–7.6441–8..6864] $\lvert\vec{z}_1(\mu_1)\rvert$=36.4280 $\lvert\vec{z}_2(\mu_1)\rvert$=12.1732 $\lvert\vec{Z}_3(\mu_1)\rvert$=12.6572	6	2.3760	0.0466

4	$\vec{z}_1(\mu_1)$=[14–7.5–23.75–27] $\vec{z}_2(\mu_1)$=[4.4025–2.3585–7.4685–8..4905] $\vec{Z}_3(\mu_1)$=[4.6178–2.4738–7.8337–8..9057] $\lvert\vec{z}_1(\mu_1)\rvert$=39.3105 $\lvert\vec{z}_2(\mu_1)\rvert$=12.3617 $\lvert\vec{Z}_3(\mu_1)\rvert$=12.9662	6.5	2.6400	0.0794
5	$\vec{z}_1(\mu_1)$=[15–8. 25.5–29] $\vec{z}_2(\mu_1)$=[4.4617–2.3796–7.5849–8..6260] $\vec{Z}_3(\mu_1)$=[4.7356–2.5256–8..0505–9.1555] $\lvert\vec{z}_1(\mu_1)\rvert$=42.1930 $\lvert\vec{z}_2(\mu_1)\rvert$=12.5502 $\lvert\vec{Z}_3(\mu_1)\rvert$=13.3206	7	2.9040	0.1122
6	$\vec{z}_1(\mu_1)$=[16–8..5–27.25–31] $\vec{z}_2(\mu_1)$=[4.5217–2.4022–7.7011–8..7609] $\vec{Z}_3(\mu_1)$=[4.8575–2.580–8..2729–9.4114] $\lvert\vec{z}_1(\mu_1)\rvert$=45.0756 $\lvert\vec{z}_2(\mu_1)\rvert$=12.7388 $\lvert\vec{Z}_3(\mu_1)\rvert$=13.6846	7.5	3.1680	0.1449

7	$\vec{z}_1(\mu_1)$=[17–9 29–33] $\vec{z}_2(\mu_1)$=[4.5824–2.4260–7.8171–8..8953] $\vec{Z}_3(\mu_1)$=[4.9813–2.6372–8..4975–9.6696] $\lvert\vec{z}_1(\mu_1)\rvert$=47.9583 $\lvert\vec{z}_2(\mu_1)\rvert$=12.9274 $\lvert\vec{Z}_3(\mu_1)\rvert$=14.0526	8.	3.4320	0.1777

TABLE 6 Numerical values of the ratio of module and appropriate coordinates of predictable points-vectors for different values of the parameters $5{,}2879 \le \mu_1 \le 8$ and $0 \le \mu_3 \le 0{,}1777$ in 4-dimensional vector space.

$\mu_1 = 5{,}2879$ $\mu_2 = 2$ $\mu_3 = 0$	$\lvert\vec{z}_1(\mu_1)\rvert / \lvert\vec{Z}_3(\mu_1)\rvert$ 2.7150	$\lvert\vec{z}_2(\mu_1)\rvert / \lvert\vec{Z}_3(\mu_1)\rvert_1$	$\lvert\vec{z}_1(\mu_1)\rvert / \lvert\vec{z}_2(\mu_1)\rvert$
	$\vec{z}_1(1)/\vec{Z}_3(1)$ 2.7150	$\vec{z}_2(1)/\vec{Z}_3(1)$ 1	$\vec{z}_1(1)/\vec{z}_2(1)$ 2.7150
	$\vec{z}_1(2)/\vec{Z}_3(2)$ 2.7150	$\vec{z}_2(2)/\vec{Z}_3(2)$ 1	$\vec{z}_1(2)/\vec{z}_2(2)$ 2.7150
	$\vec{z}_1(3)/\vec{Z}_3(3)$ 2.7150	$\vec{z}_2(3)/\vec{Z}_3(3)$ 1	$\vec{z}_1(3)/\vec{z}_2(3)$ 2.7150
	$\vec{z}_1(4)/\vec{Z}_3(4)$ 2.7150	$\vec{z}_2(4)/\vec{Z}_3(4)$ 1	$\vec{z}_1(4)/\vec{z}_2(4)$ 2.7150

	$\lvert\vec{z}_1(\mu_1)\rvert / \lvert\vec{Z}_3(\mu_1)\rvert$ 2.8299	$\lvert\vec{z}_2(\mu_1)\rvert / \lvert\vec{Z}_3(\mu_1)\rvert$ 1.0110	$\lvert\vec{z}_1(\mu_1)\rvert / \lvert\vec{z}_2(\mu_1)\rvert$ 2.7990
	$\vec{z}_1(1)/\vec{Z}_3(1)$ 2.8299	$\vec{z}_2(1)/\vec{Z}_3(1)$ 1.0110	$\vec{z}_1(1)/\vec{z}_2(1)$ 2.7990
$\mu_1 = 5{,}5$ $\mu_2 = 2{,}112$ $\mu_3 = 0{,}0139$	$\vec{z}_1(2)/\vec{Z}_3(2)$ 2.8299	$\vec{z}_2(2)/\vec{Z}_3(2)$ 1.0110	$\vec{z}_1(2)/\vec{z}_2(2)$ 2.7990
	$\vec{z}_1(3)/\vec{Z}_3(3)$ 2.8299	$\vec{z}_2(3)/\vec{Z}_3(3)$ 1.0110	$\vec{z}_1(3)/\vec{z}_2(3)$ 2.7990
	$\vec{z}_1(4)/\vec{Z}_3(4)$ 2.8299	$\vec{z}_2(4)/\vec{Z}_3(4)$ 1.0110	$\vec{z}_1(4)/\vec{z}_2(4)$ 2.7990
	$\lvert\vec{z}_1(\mu_1)\rvert / \lvert\vec{Z}_3(\mu_1)\rvert$ 2.8781	$\lvert\vec{z}_2(\mu_1)\rvert / \lvert\vec{Z}_3(\mu_1)\rvert$ 0.9618	$\lvert\vec{z}_1(\mu_1)\rvert / \lvert\vec{z}_2(\mu_1)\rvert$ 2.9925
$\mu_1 = 6$ $\mu_2 = 2{,}3760$ $\mu_3 = 0{,}0466$	$\vec{z}_1(1)/\vec{Z}_3(1)$ 2.8781	$\vec{z}_2(1)/\vec{Z}_3(1)$ 0.9618	$\vec{z}_1(1)/\vec{z}_2(1)$ 2.9925
	$\vec{z}_1(2)/\vec{Z}_3(2)$ 2.8781	$\vec{z}_2(2)/\vec{Z}_3(2)$ 0.9618	$\vec{z}_1(2)/\vec{z}_2(2)$ 2.9925
	$\vec{z}_1(3)/\vec{Z}_3(3)$ 2.8781	$\vec{z}_2(3)/\vec{Z}_3(3)$ 0.9618	$\vec{z}_1(3)/\vec{z}_2(3)$ 2.9925
	$\vec{z}_1(4)/\vec{Z}_3(4)$ 2.8781	$\vec{z}_2(4)/\vec{Z}_3(4)$ 0.9618	$\vec{z}_1(4)/\vec{z}_2(4)$ 2.9925

	$\lvert\vec{z}_1(\mu_1)\rvert/\lvert\vec{Z}_3(\mu_1)\rvert$	$\lvert\vec{z}_2(\mu_1)\rvert/\lvert\vec{Z}_3(\mu_1)\rvert$	$\lvert\vec{z}_1(\mu_1)\rvert/\lvert\vec{z}_2(\mu_1)\rvert$
	3.0318	0.9534	3.1800
	$\vec{z}_1(1)/\vec{Z}_3(1)$	$\vec{z}_2(1)/\vec{Z}_3(1)$	$\vec{z}_1(1)/\vec{z}_2(1)$
	3.0318	0.9534	3.1800
	$\vec{z}_1(2)/\vec{Z}_3(2)$	$\vec{z}_2(2)/\vec{Z}_3(2)$	$\vec{z}_1(2)/\vec{z}_2(2)$
$\mu_1 = 6,5$ $\mu_2 = 2,64$ $\mu_3 = 0,0794$	3.0318	0.9534	3.1800
	$\vec{z}_1(3)/\vec{Z}_3(3)$	$\vec{z}_2(3)/\vec{Z}_3(3)$	$\vec{z}_1(3)/\vec{z}_2(3)$
	3.0318	0.9534	3.1800
	$\vec{z}_1(4)/\vec{Z}_3(4)$	$\vec{z}_2(4)/\vec{Z}_3(4)$	$\vec{z}_1(4)/\vec{z}_2(4)$
	3.0318	0.9534	3.1800
	$\lvert\vec{z}_1(\mu_1)\rvert/\lvert\vec{Z}_3(\mu_1)\rvert$	$\lvert\vec{z}_2(\mu_1)\rvert/\lvert\vec{Z}_3(\mu_1)\rvert$	$\lvert\vec{z}_1(\mu_1)\rvert/\lvert\vec{z}_2(\mu_1)\rvert$
	3.1675	0.9422	3.3619
$\mu_1 = 7$ $\mu_2 = 2{,}904$ $\mu_3 = 0{,}1122$	$\vec{z}_1(1)/\vec{Z}_3(1)$	$\vec{z}_2(1)/\vec{Z}_3(1)$	$\vec{z}_1(1)/\vec{z}_2(1)$
	3.1675	0.9422	3.3619
	$\vec{z}_1(2)/\vec{Z}_3(2)$	$\vec{z}_2(2)/\vec{Z}_3(2)$	$\vec{z}_1(2)/\vec{z}_2(2)$
	3.1675	0.9422	3.3619
	$\vec{z}_1(3)/\vec{Z}_3(3)$	$\vec{z}_2(3)/\vec{Z}_3(3)$	$\vec{z}_1(3)/\vec{z}_2(3)$
	3.1675	0.9422	3.3619
	$\vec{z}_1(4)/\vec{Z}_3(4)$	$\vec{z}_2(4)/\vec{Z}_3(4)$	$\vec{z}_1(4)/\vec{z}_2(4)$
	3.1675	0.9422	3.3619

	$\lvert\vec{z}_1(\mu_1)\rvert / \lvert\vec{Z}_3(\mu_1)\rvert$ 3.2939	$\lvert\vec{z}_2(\mu_1)\rvert / \lvert\vec{Z}_3(\mu_1)\rvert$ 0.9309	$\lvert\vec{z}_1(\mu_1)\rvert / \lvert\vec{z}_2(\mu_1)\rvert$ 3.5385
	$\vec{z}_1(1)/\vec{Z}_3(1)$ 3.2939	$\vec{z}_2(1)/\vec{Z}_3(1)$ 0.9309	$\vec{z}_1(1)/\vec{z}_2(1)$ 3.5385
$\mu_1 = 7{,}5$ $\mu_2 = 3{,}168$ $\mu_3 = 0{,}1449$	$\vec{z}_1(2)/\vec{Z}_3(2)$ 3.2939	$\vec{z}_2(2)/\vec{Z}_3(2)$ 0.9309	$\vec{z}_1(2)/\vec{z}_2(2)$ 3.5385
	$\vec{z}_1(3)/\vec{Z}_3(3)$ 3.2939	$\vec{z}_2(3)/\vec{Z}_3(3)$ 0.9309	$\vec{z}_1(3)/\vec{z}_2(3)$ 3.5385
	3.2939	$\vec{z}_2(4)/\vec{Z}_3(4)$ 0.9309	$\vec{z}_1(4)/\vec{z}_2(4)$ 3.5385
	$\lvert\vec{z}_1(\mu_1)\rvert / \lvert\vec{Z}_3(\mu_1)\rvert$ 3.4128	$\lvert\vec{z}_2(\mu_1)\rvert / \lvert\vec{Z}_3(\mu_1)\rvert$ 0.9199	$\lvert\vec{z}_1(\mu_1)\rvert / \lvert\vec{z}_2(\mu_1)\rvert$ 3.7098
$\mu_1 = 8$ $\mu_2 = 3{,}432$ $\mu_3 = 0{,}1777$	$\vec{z}_1(1)/\vec{Z}_3(1)$ 3.4128	$\vec{z}_2(1)/\vec{Z}_3(1)$ 0.9199	$\vec{z}_1(1)/\vec{z}_2(1)$ 3.7098
	$\vec{z}_1(2)/\vec{Z}_3(2)$ 3.4128	$\vec{z}_2(2)/\vec{Z}_3(2)$ 0.9199	$\vec{z}_1(2)/\vec{z}_2(2)$ 3.7098
	$\vec{z}_1(3)/\vec{Z}_3(3)$ 3.4128	$\vec{z}_2(3)/\vec{Z}_3(3)$ 0.9199	$\vec{z}_1(3)/\vec{z}_2(3)$ 3.7098
	$\vec{z}_1(4)/\vec{Z}_3(4)$ 3.4128	$\vec{z}_2(4)/\vec{Z}_3(4)$ 0.9199	$\vec{z}_1(4)/\vec{z}_2(4)$ 3.7098

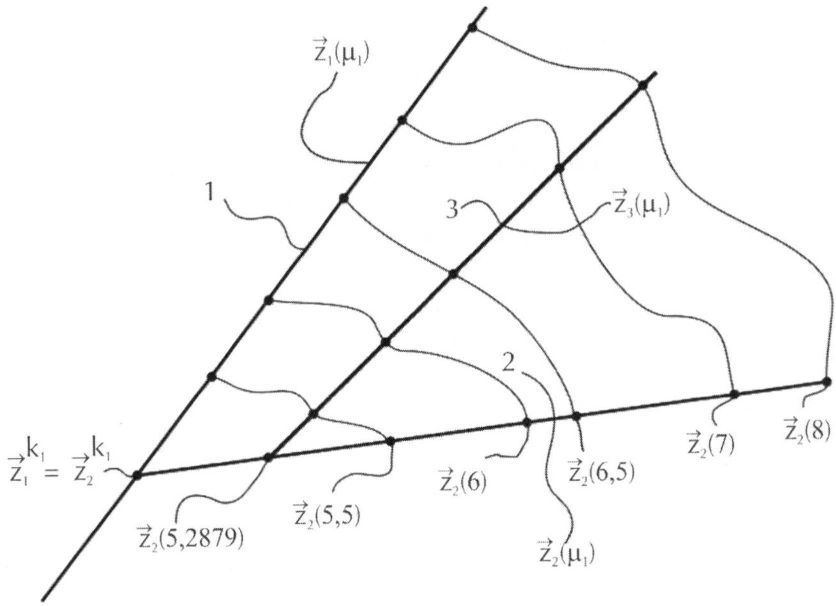

FIGURE 2 The graph of numerical values of module and appropriate coordinates of predictable points-vectors for the values of the parameters μ_1 and μ_3, changing in the interval $5{,}2879 \le \mu_1 \le 8$ and $0 \le \mu_3 \le 0{,}1777$, calculated by different criteria in 4-dimensional space.

Here, for any value of the arbitrary parameter μ_1 changing in the interval $\mu_1^{k_2} \le \mu_1 \le \mu^*$ we have the appropriate numerical values $\vec{z}_1(\mu_1)$, $\vec{z}_2(\mu_1)$, $\vec{Z}_3(\mu_1)$, $|\vec{z}_1(\mu_1)|$, $|\vec{z}_2(\mu_1)|$, $|\vec{Z}_3(\mu_1)|$.

For visually, as an example take the values of the parameter $\mu_1 = 6$.

Take into attention the denotation of appropriate ratios of the coordinates of the vectors $\vec{z}_1(\mu_1)$ $\vec{z}_2(\mu_1)$ $\vec{Z}_3(\mu_1)$ for (*i*: 1, 2, 3, 4) in the form:

$$n_{1i} = \frac{\vec{z}_1(i)}{\vec{Z}_3(i)} = \frac{x_{1i}}{X_{3i}}, \; n_{2i} = \frac{\vec{z}_2(i)}{\vec{Z}_3(i)} = \frac{x_{2i}}{X_{3i}},$$

$$n_{3i} = \frac{\vec{z}_1(i)}{\vec{z}_2(i)} = \frac{x_{1i}}{x_{2i}} \tag{35}$$

$$n_4 = \frac{|\vec{z}_1(\mu_1)|}{|\vec{Z}_3(\mu_1)|}, \; n_5 = \frac{|\vec{z}_2(\mu_1)|}{|\vec{Z}_3(\mu_1)|},$$

$$n_6 = \frac{|\vec{z}_1(\mu_1)|}{|\vec{z}_2(\mu_1)|} \tag{36}$$

According to Eqs. (35)–(36) and Tables 5 and 6 numerically establish the ratios of the coordinates of the vectors $\vec{z}_1(\mu_1)$ and $\vec{z}_2(\mu_1)$, i.e., x_{1i} and x_{2i} from the appropriate coordinates of the prediction function $\vec{Z}_3(\mu_1)$ with regard to uncertainty factors, in the form:

$$n_1 = n_2 = n_3 = 2{,}8781 \tag{37}$$

$$n_2 = n_2 = n_3 = 0{,}9618 \tag{38}$$

$$n_3 = n_2 = n_3 = 2{,}9925 \tag{39}$$

$$n_4 = 2{,}8781 \,,\; n_5 = 0{,}9618 \,,\; n_6 = 2{,}9925 \tag{40}$$

- Numerical value Eq. (37) shows that the values of coordinates of prediction variables calculated by linear criterion, is 2.8781 times higher than appropriate prediction coordinates calculated according to vectors-function with regard to uncertainty factors influence;
- numerical value Eq. (38) shows that the values of the coordinates of prediction variables calculated by the means of the second piecewise-linear vector function composes 0.9618 part from the values of appropriate prediction coordinates calculated according to vector-function with regard to uncertainty factors influence;
- numerical value Eq. (39) shows that the values of coordinates of prediction variables calculates by the linear criterion is 2.9925 times higher than the appropriate prediction coordinates calculated by means of the second piecewise-linear vector-function;
- numerical value Eq. (40) shows percentage ratio of total indices of vector-function, i.e., by the module of vectors $|\vec{z}_1(\mu_1)|$, $|\vec{z}_2(\mu_1)|$, $|\vec{Z}_3(\mu_1)|$, calculated by different criteria.

By means of numerical data of Table 4, it is easy to establish the dependence of coordinates of prediction vector-function depending on the parameter μ_1, i.e., $\vec{z}_1(i) \sim \mu_1$, $\vec{z}_2(i) \sim \mu_1$ and $\vec{Z}_3(i) \sim \mu_1$.

5.5 DEVELOPMENT OF SOFTWARE FOR COMPUTER MODELING AND MULTIVARIANT PREDICTION OF ECONOMIC EVENT AT UNCERTAINTY CONDITIONS ON THE BASE OF *N*-COMPONENT PIECEWISE-LINEAR ECONOMIC-MATHEMATICAL MODELS ON A PLANE

In Chapters 2 and 3, we developed a general theory on construction of piecewise-linear models and gave a method of multivariant prediction in m-dimensional vector space, in Chapter 5 and Sections 5.1.4 suggested a packet of numerical programs worked out in the *Matlab* language.

In Chapter 4, the represented theory was applied for the case of 2-dimensional economic process, and the statement was given in coordinate form.

Here, geometric interpretation of the introduced unaccounted parameter λ_m and unaccounted factors influence function $\omega_n(t, \lambda_n)$ was represented in detail, the criteria of economic process prediction and its control at uncertainty conditions on a plane was determined, a mechanism of conduction of numerical calculations on the given method was shown [5, 6, 9, 17, 18].

The goal of the present section is to develop a software for computer modeling of piecewise-linear economic-mathematical models with regard to unaccounted factors influence on a plane and to approve the efficiency of the given program on examples.

Follow the algorithm of modeling of the given class of economic-mathematical models.

5.5.1 ACTIONS ALGORITHM FOR COMPUTER SIMULATION BY CONSTRUCTING N-COMPONENT PIECEWISE-LINEAR ECONOMIC-MATHEMATICAL MODELS ON A PLANE

Let a system of statistical points $\{t_i, y_i(t)\}$ be given on a plane. By the known methods we approximate this system of points in the form of piece-wise-linear straight lines in the segments of the form [5, 6, 9, 17, 18]:

$$y_n(t) = a_n t + b_n \text{ for } (t_n \le t \le t_{n+1}). \quad (41)$$

Let the Eq. (41) and conjugation points between the appropriate straight lines $\{t_n, y_n(t_n)\}$ be given.

The goal of the investigation is to develop a computer algorithm for calculations of the following expressions and functions by the given Eq. (41) and conjugation points $\{t_n, y_n(t_n)\}$:

$$y_n(t_n) = a_n t_n + b_n$$

$$y_n(t_{n+1}) = a_n t_{n+1} + b_n$$

$$E_n = \frac{y_n(t_{n+1}) - y_n(t_n)}{t_{n+1} - t_n}$$

$$A_n = \prod_{k=2}^{n-1} B_k = \prod_{k=1}^{n-1} [1 - \omega_k(t_{k+1}, \lambda_k)]$$

$$A_1 = \prod_{k=2}^{0} [1 - \omega_k(t_{k+1}, \lambda_k)] = 1$$

$$A_2 = \prod_{k=2}^{1} [1 - \omega_k(t_{k+1}, \lambda_k)] = 1$$

$$\lambda_n = \frac{A_n - {E_n}/{E_1}}{A_n}$$

$$\omega_n(t, \lambda_n) = \lambda_n(1 - \frac{t_n}{t}) \text{ for } t \geq t_n$$

$$\omega_n(t_{n+1}, \lambda_n) = \lambda_n(1 - \frac{t_n}{t_{n+1}})$$

$$B_k = 1 - \omega_k(t_{k+1}, \lambda_k)$$

$$y_n(t, \lambda_n) = E_1 t[1 - \omega_n(t, \lambda_n)] \cdot A_n + y_1(t_1) - E_1 t_1 \tag{42}$$

Introduce the following denotation:

$$a_n \to a(n),\ b_n \to b(n),\ t_n \to t(n),\ y_n(t_n) \to yh(n),$$

$$y_n(t_{n+1}) \to ye(n),\ E_n \to E(n),\ A_n \to A(n),\ \lambda_n \to La(n),$$

$$\omega_n(t_{n+1}, \lambda_n) \to W(n),\ B_k \to B(n),$$

$$\begin{cases} E_1(1 - \lambda_n) \cdot A_n \to alfa(n) \\ E_1 \cdot \lambda_n \cdot t_n \cdot A_n + y_1(t_1) - E_1 t_1 \to betta(n) \end{cases}$$

$$y_n(t, \lambda_n) = alfa(n) \cdot T + betta(n) \tag{43}$$

Using the introduced denotation, based around the Matlab program compose a program for numerical construction of n-component piecewise-linear economic-mathematical models with regard to unaccounted factors influence, in the following form:

```
M=N;
A = zeros(1, M+1);
B = zeros(1, M);
La = zeros(1, M);
a = zeros(1, M);
b = zeros(1, M);
t = zeros(1, M+1);
E = zeros(1, M);
W = zeros(1, M);
ye = zeros(1, M);
yh = zeros(1, M);
Y= zeros(1, M);
A(1)=1;
A(2)=1;
La(1)=0;
B(1)=1;
for n=1:M
yh(n)=a(n)*t(n)+b(n);
ye(n)=a(n)*t(n+1)+b(n);
E(n)=(ye(n)-yh(n))/(t(n+1)–t(n));
end
for n=2:M
A(n)=A(n–1)*B(n–1);
La(n)=(A(n)–(E(n)/E(1)))/A(n);
W(n)=La(n)*(1–t(n)/t(n+1));
```

```
B(n)=1–W(n);
End
A(M+1)=A(M)*B(M);
for n=1:M
alfa(n)=E(1)*(1-la(n))*A(n);
betta(n)=E(1)*La(n)*t(n)*A(n)+yh(1)-E(1)*t(1);
end
S=sym('E(1)*T*(1-la(n)*(1-(t(n)/T)))*A(n)+yh(1)–
-E(1)*t(1)');
pc=collect(S, 'T');
pretty(pc);
alfa
betta
```
(44)

The suggested numerical program (44) allows to determine numerically the points of any piecewise-linear straight line $y_n(t,\lambda_n)$ depending on the first piecewise-linear straight line $y_1(t,\lambda_1)$ and all kinds of unaccounted parameters influence functions $\omega_n(t,\lambda_n)$ met on all (n-1) preceding time intervals of economic event, i.e., depending on $\omega_2(t,\lambda_2)$, $\omega_3(t,\lambda_3)$, $\omega_4(t,\lambda_4)$, $\omega_5(t,\lambda_5)$, …., $\omega_{n-1}(t,\lambda_{n-1})$ unaccounted parameters influence functions [4–6, 9, 17, 18].

Example. As an example consider the table of statistical data

a=[4, 2, 1.2, 0.5, 0.3332];

b=[–2, 4, 9.6, 17.9967, 21];

t=[1, 3, 7, 12, 18, 21];

that represents itself as five-component model given on time interval $0 \le t \le t_6$.

Here the task of the investigation is:

– to represent the given statistical table in the form of a unique 5 piece-wise-linear functions so that they were represented depending on the form of the first piecewise-linear function $y_1(t,\lambda_1)$ and all unaccounted parameters influence functions met on all 5 preceding time intervals of economic event, i.e., $\omega_2(t,\lambda_2)$, $\omega_3(t,\lambda_3)$, $\omega_4(t,\lambda_4)$, $\omega_5(t,\lambda_5)$.

As the program is completely set up in the window "Tditor", we press the window "Run" on the plug board.

As a result, the Fig. 3 of the constructed model automatically emerges in the window "Figure" and necessary mathematical calculations in the window "Command Window"

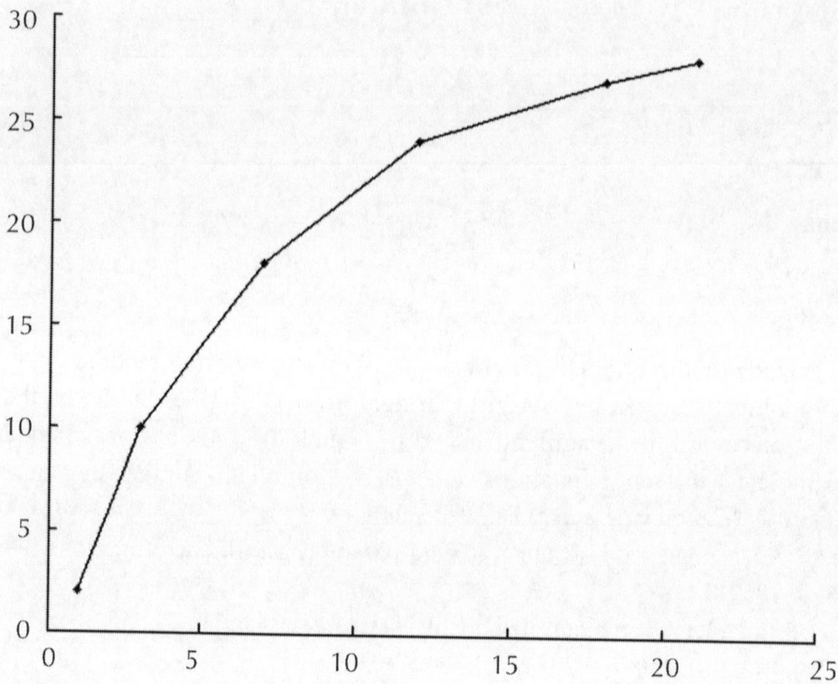

FIGURE 3 Figure in the window "Figure."

a = 4.0000–2.0000–1.2000–0.5000–0.3332

b = −2.0000–4.0000–9.6000–17.9967–21.0000

yh = 2.0000–10.0000–18.0000–23.9967–26.9976

ye = 10.0000–18.0000–24.0000–26.9967–27.9972

E = 4.0000–2.0000–1.2000–0.5000–0.3332

A = 1.0000–1.0000–0.7143–0.5417–0.4028–0.3571

La = 0–0.5000–0.5800–0.7692–0.7932

W = 0–0.2857–0.2417–0.2564–0.1133

B = 1.0000–0.7143–0.7583–0.7436–0.8867

E(1) (1 – La(n)) A(n) T + E(1) La(n) t(n) A(n) +yh(1) – E(1) t(1)

alfa = 4.0000–2.0000–1.2000–0.5000–0.3332

betta = –2.0000–4.0000–9.6000–18.0000–21.0024.

These number data are perfectly coincide with the results obtained in Section 4.4 that were represented by Fig. 3 and Table 1 of Chapter 4 [4, 9, 17, 18].

5.1.2 ALGORITHM OF MULTIVARIANT COMPUTER MODELING OF PREDICTION VARIABLES OF ECONOMIC EVENT BY MEANS OF *N*-COMPONENT PIECEWISE-LINEAR ECONOMIC-MATHEMATICAL MODELS ON A PLANE

In the preceding section we developed a special software for computer modeling of numerical construction of piecewise-linear economic-mathematical models with regard to unaccounted factors influence on a plane [4–6, 9, 17, 18].

In this section, by means of the Matlab program, we suggest a software for a multivariant computer modeling of prediction variables of economic event by piecewise-linear economic-mathematical models with regard to unaccounted factor influence on a plane.

And on a concrete example we show an action algorithm on the given program.

Let on a plane be given a system of statistical points $\{t_i, y_i(t)\}$. We approximate this system of points by the know methods in the form of n-component piecewise-linear straight lines in the segments of the form [4–6, 9, 17, 18].

$$y_n(t) = a_n t + b_n \text{ for } (t_n \le t \le t_{n+1}) \tag{45}$$

We assume that these Eq. (45) and the conjugation points between the appropriate straight lines $\{t_n, y_n(t_n)\}$ are given.

The goal of the investigation is to develop a computer algorithm of calculations of the following expressions and functions by the given Eq. (45) and conjugation points $\{t_n, y_n(t_n)\}$:

$$y_n(t_n) = a_n t_n + b_n$$

$$y_n(t_{n+1}) = a_n t_{n+1} + b_n$$

$$E_n = \frac{y_n(t_{n+1}) - y_n(t_n)}{t_{n+1} - t_n}$$

$$A_n = \prod_{k=2}^{n-1} B_k = \prod_{k=1}^{n-1} [1 - \omega_k(t_{k+1}, \lambda_k)].$$

Here

$$A_1 = \prod_{k=2}^{0} [1 - \omega_k(t_{k+1}, \lambda_k)] = 1$$

$$A_2 = \prod_{k=2}^{1} [1 - \omega_k(t_{k+1}, \lambda_k)] = 1$$

$$\lambda_n = \frac{A_n - E_n / E_1}{A_n}$$

$$\omega_n(t,\lambda_n)=\lambda_n(1-\frac{t_n}{t})\text{ for }t\geq t_n$$

$$\omega_n(t_{n+1},\lambda_n)=\lambda_n(1-\frac{t_n}{t_{n+1}})\quad B_k=1-\omega_k(t_{k+1},\lambda_k)$$

$$y_n(t,\lambda_n)=E_1t[1-\omega_n(t,\lambda_n)]\cdot A_n+y_1(t_1)-E_1t_1$$

$$y_{n+1}(t,\lambda_{n+1})=E_1t[1-\omega_m(t,\lambda_m)]\cdot A_{N+1}+y_1(t_1)-E_1t_1$$

$$\lambda_m=\frac{\prod_{\alpha=2}^{m-1\geq 2}B_\alpha-E_m/E_1}{A_n}$$

$$\omega_m(t,\lambda_m)=\lambda_m(1-\frac{t_{N+1}}{t})$$

$$A_{N+1}=\prod_{k=2}^{N}B_k=\prod_{k=2}^{N}[1-\omega_k(t_{k+1},\lambda_k)] \tag{46}$$

Introduce the denotation:

$a_n\rightarrow a(n)$, $b_n\rightarrow b(n)$, $t_n\rightarrow t(n)$, $y_n(t_n)\rightarrow yh(n)$

$y_n(t_{n+1})\rightarrow ye(n)$, $E_n\rightarrow E(n)$, $A_n\rightarrow A(n)$, $\lambda_n\rightarrow La(n)$

$\omega_n(t_{n+1},\lambda_n)\rightarrow W(n)$, $B_k\rightarrow B(n)$,

$$E_1(1-\lambda_n)\cdot A_n \to alfa(n)$$

$$E_1\cdot\lambda_n\cdot t_n\cdot A_n + y_1(t_1) - E_1 t_1 \to betta(n)$$

$$y_n(t,\lambda_n) = alfa(n)\cdot T + betta(n)$$

$$E_1(1-\lambda_m)\cdot A_{N+1} \to alfa\ NEW(m),$$

$$E_1\cdot\lambda_m\cdot t_{n+1}\cdot A_{N+1} + y_1(t_1) - E_1 t_1 \to betta\ NEW(m)$$

$$y_{N+1}(t,\lambda_{N+1},\lambda_m) = alfa\ NEW(m)\cdot Td + betta\ NEW(m)$$

Td=[2:N]. (47)

Using the introduced denotation (47), compose an algorithm of actions by composing this program, in the following form:

```
M=M*;

A = zeros(1, M+1);

B = zeros(1, M);

La = zeros(1, M);

a = zeros(1, M);

b = zeros(1, M);

t = zeros(1, M+1);

E = zeros(1, M);
```

```
W = zeros(1, M);
e = zeros(1, M);
yh = zeros(1, M);
P= zeros(1, M);
A(1)=1;
A(2)=1;
La(1)=0;
B(1)=1;
for n=1:M
yh(n)=a(n)*t(n)+b(n);
ye(n)=a(n)*t(n+1)+b(n);
E(n)=(ye(n)–yh(n))/(t(n+1)–t(n));
end
for n=2:M
A(n)=A(n-1)*B(n-1);
La(n)=(A(n)–(E(n)/E(1)))/A(n);
W(n)=La(n)*(1–t(n)/t(n+1));
B(n)=1–W(n);
end
A(M+1)=A(M)*B(M);
a
```

```
b
t
yh
ye
E
A
La
W
B
for n=1:M
alfa(n)=E(1)*(1-la(n))*A(n);
betta(n)=E(1)*La(n)*t(n)*A(n)+yh(1)–E(1)*t(1);
end
S = sym('E(1)*T*(1–la(n)*(1–(t(n)/T)))*A(n)+yh(1)–E(1)*t(1)');
pc=collect(S, 'T');
pretty(pc);
alfa
betta
for m=1:M
alfaNEW(m)=E(1)*(1–la(m))*A(M+1);
bettaNEW(m)=E(1)*La(m)*t(M+1)*A(M+1)+yh(1)–E(1)*t(1);
```

```
end

alfaNEW

bettaNEW

Td=24;

for n=1:M

line([t(n) t(n+1)], [yh(n)ye(n)], 'Marker,' '.,' 'LineStyle,' '–');

end

for n=two:M

P(n)=alfaNEW(n)*Td+bettaNEW(n);line([t(M+1)Td],

[ye(M) P(n)], 'Marker,' '.,' 'LineStyle,' '–');

end

Td

P
```

Thus, giving the statistical data a_n, b_n, and also the conjugation points $(t_n, y_n(t_n))$ by means of the developed algorithm we constructed a unique n-component piecewise-linear economic-mathematical, model of economic event, that defines the form of influence of unaccounted parameters λ_n and the function, $\omega_n(\lambda_n, \alpha_{n-1,n})$, and also defining the numerical value of the points of multivariant prediction of economic event on the subsequent stage [4–6, 9, 17, 18].

Example. As an example consider the table of statistical data:

a=[4–2 1.2–0.5–0.3332];

b=[–2–4 9.6–17.9967–21];

t=[1–3 7–12–18–21];

that represents itself as a five-component model given on time interval $0 \le t \le t_6$.

The task of the investigation is:

- to represent the given statistical table in the form of a unique, five piecewise-linear function such that it was represented depending on the form of the first piecewise-linear function $\vec{z}_1$ and all unaccounted parameters influence function met on all 5 preceding time intervals of economic event, i.e., $\omega_2(t,\lambda_2)$, $\omega_3(t,\lambda_3)$, $\omega_4(t,\lambda_4)$, $\omega_5(t,\lambda_5)$.
- secondly, by the characteristics of the preceding five piecewise-linear functions given on the common time interval (t_1, t_6), to construct the sixth piecewise-linear function on the subsequent time interval (t_1, t_7). Applying the above-developed algorithm to the problem under investigation we follow the following continuation:

As the program is completely set up in the window "Editor" we press the window "Run" on the plugboard.

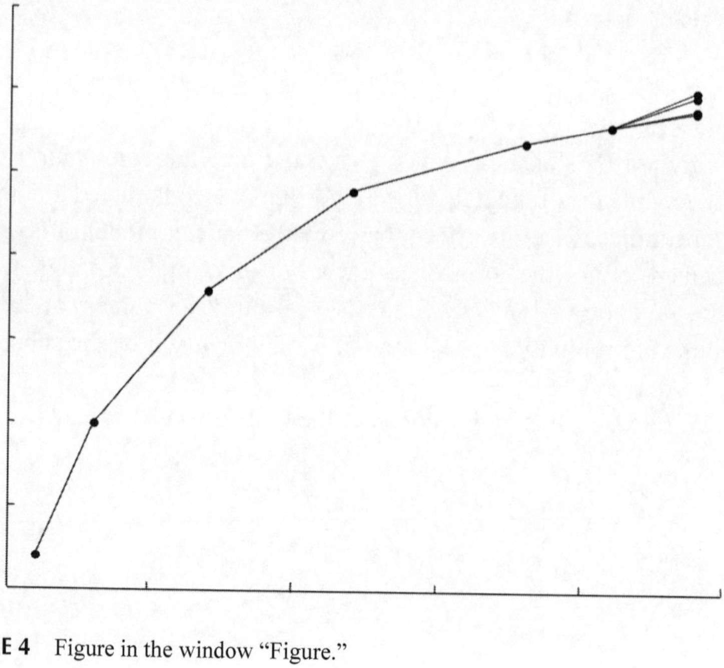

FIGURE 4 Figure in the window "Figure."

As a result, the Fig. 4 of the constructed model automatically emerges in the window "Figure", and necessary mathematical calculations in the window "Command Window."

a = 4.0000–2.0000–1.2000–0.5000–0.3332

b = −2.0000–4.0000–9.6000–17.9967–21.0000

t = 1–3 7–12–18–21

yh = 2.0000–10.0000–18.0000–23.9967–26.9976

ye = 10.0000–18.0000–24.0000–26.9967–27.9972

E = 4.0000–2.0000–1.2000–0.5000–0.3332

A = 1.0000–1.0000–0.7143–0.5417–0.4028–0.3571

La = 0–0.5000–0.5800–0.7692–0.7932

W = 0–0.2857–0.2417–0.2564–0.1133

B = 1.0000–0.7143–0.7583–0.7436–0.8867

E(1) (1 – La(n)) A(n) T + E(1) La(n) t(n) A(n) + yh(1) – E(1) t(1)

alfa = 4.0000–2.0000–1.2000–0.5000–0.3332

betta = −2.0000–4.0000–9.6000–18.0000–21.0024

alfaNEW = 1.4286–0.7143–0.6000–0.3297–0.2954

bettaNEW = −2.0000–12.9998–15.3998–21.0766–21.7953

Td = 24

P = 0–30.1424–29.7996–28.9886–28.8859

a = 4.0000–2.0000–1.2000–0.5000–0.3332

b = −2.0000–4.0000–9.6000–17.9967–21.0000

yh = 2.0000–10.0000–18.0000–23.9967–26.9976

ye = 10.0000–18.0000–24.0000–26.9967–27.9972

E = 4.0000–2.0000–1.2000–0.5000–0.3332

A = 1.0000–1.0000–0.7143–0.5417–0.4028–0.3571

La = 0–0.5000–0.5800–0.7692–0.7932

W = 0–0.2857–0.2417–0.2564–0.1133

B = 1.0000–0.7143–0.7583–0.7436–0.8867

E(1) (1 – La(n)) A(n) T + E(1) La(n) t(n) A(n) + yh(1) – E(1) t(1)

alfa = 4.0000–2.0000–1.2000–0.5000–0.3332

betta = −2.0000–4.0000–9.6000–18.0000–21.0024

E(1) (1 – La(n))A(n) T + E(1)La(n) t(n +1) A(n) + yh(1) – E(1) t(1)

alfaNEW = 1.4286–0.7143–0.6000–0.3297–0.2954

bettaNEW = −2.0000–12.9998–15.3998–21.0766–21.7953

a = 4.0000–2.0000–1.2000–0.5000–0.3332

b = −2.0000–4.0000–9.6000–17.9967–21.0000

t = one–three seven–12–18–21

yh = 2.0000–10.0000–18.0000–23.9967–26.9976

ye = 10.0000–18.0000–24.0000–26.9967–27.9972

E = 4.0000–2.0000–1.2000–0.5000–0.3332

A = 1.0000–1.0000–0.7143–0.5417–0.4028–0.3571

La = 0–0.5000–0.5800–0.7692–0.7932

W = 0–0.2857–0.2417–0.2564–0.1133

B = 1.0000–0.7143–0.7583–0.7436–0.8867

E(1) (1 – La(n)) A(n) T + E(1) La(n) t(n) A(n) + yh(1) – E(1) t(1)

alfa = 4.0000–2.0000–1.2000–0.5000–0.3332

betta = -2.0000–4.0000–9.6000–18.0000–21.0024

alfaNEW = 1.4286–0.7143–0.6000–0.3297–0.2954

bettaNEW = −2.0000–12.9998–15.3998–21.0766–21.7953

Td = 24

P = 0–30.1424–29.7996–28.9886–28.8859

This completely corresponds to graphical representation of the earlier developed plane piecewise-linear economic-mathematical model with regard to unaccounted factors influence (Fig. 3 – construction of a piecewise-linear function with upwards convexity and Table 1 of Chapter 4), and also (Fig. 4 – establishment of the range of the predictable function with upwards convexity and Table 2 of Chapter 4) [4, 9, 17–24].

KEYWORDS

- **existence theorem**
- ***m*-dimensional vector space**
- ***Matlab* program**
- **multivariant prediction of economic state**
- ***n*-component piecewise-linear economic-mathematical model**

REFERENCES

1. Aliyev Azad, G. *Theoretical aspects of the problem of analysis of risks of investigation projects in oilgas recovery industry.* Izv. of NASA, Ser. of human and social sciences (economics), Baku, **2009,** *1*, 97–104.
2. Aliyev Azad, G. *Definition of optimal plan of equipment in oil-gas industry by means of dynamical economic-mathematical programming.* Izvstiya NAS of Azerbaijan, Ser. human and social sciences (economics). Baku, **2009,** *4*, 37–42.
3. Aliyev Azad, G. *Development of software for computer modeling of 2-component piecewise-linear economic-mathematical model with regard to uncertainty factors influence in a 4-dimensional vector space.* Vestnik KhGAP, Khabarovsk, **2010,** *2(47)*,43–56.
4. Aliyev Azad, G. *Economical and mathematical models subject to complete information.* Ed. Lap Lambert Akademic Publishing. ISBN-978–3-659–30998–4, Berlin, **2013,** 316.
5. Aliyev Azad, G. *Economic-mathematical methods and models in uncertainty conditions in finite-dimensional vector space.* Ed. NAS of Azerbaijan "Information technologies." ISBN 978–9952–434–10–1, Baku, **2009,** 220.
6. Aliyev Azad, G. *Prospecting's of software development for computer modeling of economic event by means of piecewise-linear economic-mathematical models with regard to uncertainty factors in 3-dimensional vector space.* Proceedings of the III International

scientific-practical conference "Youth and science: reality and future," Natural and applied problems. Nevinnomisk, **2010,** *5*, 338–343.

7. Aliyev Azad, G. *Development of software for numerical construction of piecewise-linear economic-mathematical models with regard to unaccounted factors influence on a plane.* "Issues of economics science" "Sputnik" publishing. ISSN 1728–8878, Moscow, **2009,** *5*, 106–112.
8. Aliyev Azad, G. *Software development for computer modeling of 2-component piece-wise-linear economic-mathematical model with regard to uncertainty factors influence in m-dimensional vector space.* "Natural and technical sciences" "Sputnik" publishing.; ISSN 1684–2626, Moscow, **2010,** *2(46),* 510–521.
9. Aliyev Azad, G. *Theoretical bases of economic-mathematical simulation at uncertainty conditions*, National Academy of Sciences of Azerbaijan, Baku, "Information Technologies," ISBN 995221056–9, **2011,** 338.
10. Aliyev Azad, G. *Development of software for computer modeling of economic event prediction by means of piecewise-linear economic-mathematical models with regard to unaccounted factor influence on a plate.* "Economic, Statistic and Informatic," Vestnik of UMO: moscow, **2009,** *4*, 139–144.
11. Aliyev Azad, G. *Some aspects of computer modeling of economic events in uncertainty conditions in 3-dimensional vector space.* Collection of papers of the XVIII International Conference "Mathematics, Economics, and Education of the VI International Symposium "Fourier series and their applications," Rostov-na-donu, **2010,** 156–157.
12. Aliyev Azad, G. *Software development for computer modeling of 2-component piece-wise economic-mathematical model with regard to uncertainty factors influence in 3-dimensional vector space.* Publishing "Economics sciences," Moscow, **2010,** *3*, 249–256.
13. Aliyev Azad, G. *Software development for multivariant prediction of economic event in uncertainty conditions on the base of 2-component piecewise-linear economic-mathematical models in 4-dimensional vector space.* Publishing Vestnik of KhGAP, Khabarovsk, **2010,** 6*(51),* 30–42.
14. Aliyev Azad, G. *Software development for multivariant prediction of economic event in uncertainty conditions on the base of 2-component piecewise-linear economic-mathematical model in m-dimensional vector space.* Publishing *"Economics sciences,"* Moscow, **2010,** *5(66),* 236–245.
15. Aliyev Azad, G. *Software development for multivariant prediction of economic event in uncertainty conditions on the base of 2-dimensional piecewise-linear economic-mathematical model in 3-dimensional vector space.* Publishing "Economic, Statistic and Informatic" Vestnik of UMO: Moscow, **2010,** *3*, 127–136.
16. Aliyev Azad A. *Prospects of development of the software support for the computer simulation of the forecast of economic event using piecewise-linear economic-mathematical models in view of factors of uncertainty in 3-dimensional vector space.* PCI, The third international conference "Problem of cybernetics and informatics," ISBN 978-9952-453-33-1, Baku, **2010,** *3,* 185–188.
17. Aliyev Azad, G. *Development of dynamical model for economic process and its application in the field of industry of Azerbaijan.* Thesis for PHD, Baku, **2001,** 167.
18. Aliyev Azad, G. *Economic-mathematical methods and models with regard to incomplete information.* Ed. "Elm," ISBN 5-8066-1487-5, Baku, **2002,** 288.

19. Aliyev Azad, G.; Ecer, F. *Tam olmayan bilqiler durumunda iktisadi matematik metodlar ve modeler,* Ed. NUI of Turkiye, ISBN 975–8062–1802, Nigde, 2004, 223.
20. Aliyev Azad, G. *Search of decisions in fuzzy conditions*. Izv. Visshikh tekhnicheskikh uchebnikh zavedeniy Azerbayjan. ASOA, ISSN 1609–1620, Baku, **2006,** *5(45)*, 66–71.
21. Aliyev Azad, G. *Numerical calculation of 2-component piecewise-linear economic-mathematical model in 3-dimensional vector space.* Izvestia NAS of Azerbaijan, Baku, **2008,** *1*, 96–102.
22. Aliyev Azad, G. *Numerical calculation of economic event prediction with regard to uncertainty factor in 3-dimensional vector space on the base of 3-component piece-wise-linear economic-mathematical model.* Collection of papers of the Institute of Economics NAS of Azerbaijan, Baku, **2008,** 129–143.
23. Aliyev Azad, G. *Numerical calculation of economic event prediction with regard to uncertainty factor in 3-dimensional vector space on the base of 2-component piece-wise-linear economic-mathematical model.* Collection of papers of the Institute of Economics of NAS of Azerbaijan (Problems of National Economics), ISBN 5-8066-1711-3, Baku, **2008,** *1*, 142–156.
24. Aliyev Azad, G. *Numerical calculation of 3-component piecewise-linear economic-mathematical model with regard to uncertainty factor in 3-dimensional vector space.* Izvestiya NAS of Azerbaijan, Ser. of human and social sciences (economics), Baku, **2008,** *3*, 86–100.

INDEX

G

U